Hybrid Intelligence for Smart Grid Systems

Advances in Intelligent Decision-Making, Systems Engineering, and Project Management

This new book series will report the latest research and developments in the field of information technology, engineering and manufacturing, construction, consulting, healthcare, military applications, production, networks, traffic management, crisis response, human interfaces, and other related and applied fields. It will cover all project types, such as organizational development, strategy, product development, engineer-to-order manufacturing, infrastructure and systems delivery, and industries and industry-sectors where projects take place, such as professional services, and the public sector including international development and cooperation etc. This new series will publish research on all fields of information technology, engineering, and manufacturing including the growth and testing of new computational methods, the management and analysis of different types of data, and the implementation of novel engineering applications in all areas of information technology and engineering. It will also publish on inventive treatment methodologies, diagnosis tools and techniques, and the best practices for managers, practitioners, and consultants in a wide range of organizations and fields including police, defense, procurement, communications, transport, management, electrical, electronic, aerospace, requirements.

Blockchain Technology for Data Privacy Management
Edited by Sudhir Kumar Sharma, Bharat Bhushan, Aditya Khamparia, Parma Nand Astya, and Narayan C. Debnath

Smart Sensor Networks Using AI for Industry 4.0
Applications and New Opportunities
Edited by Soumya Ranjan Nayak, Biswa Mohan Sahoo, Muthukumaran Malarvel, and Jibitesh Mishra

Hybrid Intelligence for Smart Grid Systems
Edited by Seelam VSV Prabhu Deva Kumar, Shyam Akashe, Hee-Je Kim, and Chinmay Chakrabarty

For more information about this series, please visit: https://www.routledge.com/Advances-in-Intelligent-Decision-Making-Systems-Engineering-and-Project-Management/book-series/CRCAIDMSEPM

Hybrid Intelligence for Smart Grid Systems

Edited by
Seelam VSV Prabhu Deva Kumar, Shyam Akashe,
Hee-Je Kim, and Chinmay Chakraborty

CRC Press
Taylor & Francis Group
Boca Raton London New York

CRC Press is an imprint of the
Taylor & Francis Group, an **informa** business

First edition published 2022
by CRC Press
6000 Broken Sound Parkway NW, Suite 300, Boca Raton, FL 33487-2742

and by CRC Press
2 Park Square, Milton Park, Abingdon, Oxon, OX14 4RN

© 2022 selection and editorial matter, Seelam VSV Prabhu Deva Kumar, Shyam Akashe, Hee-Je Kim, and Chinmay Chakraborty; individual chapters, the contributors

CRC Press is an imprint of Taylor & Francis Group, LLC

Library of Congress Cataloging-in-Publication Data
Names: Kumar, Savitri V., 1944- editor.
Title: Hybrid intelligence for smart grid systems / edited by Seelam VSV
Prabhu Deva Kumar, Shyam Akashe, Hee-Je Kim and Chinmay Chakraborty.
Description: First edition. | Boca Raton, FL : CRC Press, 2022. | Series:
Advances in intelligent decision-making, systems engineering, and
project management | Includes bibliographical references and index.
Identifiers: LCCN 2021020840 (print) | LCCN 2021020841 (ebook) |
ISBN 9780367699093 (hbk) | ISBN 9780367699109 (pbk) |
ISBN 9781003143802 (ebk)
Subjects: LCSH: Smart power grids.
Classification: LCC TK3105 .H93 2022 (print) | LCC TK3105 (ebook) | DDC
621.31—dc23
LC record available at https://lccn.loc.gov/2021020840
LC ebook record available at https://lccn.loc.gov/2021020841

ISBN: 978-0-367-69909-3 (hbk)
ISBN: 978-0-367-69910-9 (pbk)
ISBN: 978-1-003-14380-2 (ebk)

DOI: 10.1201/9781003143802

Typeset in Times
by codeMantra

Contents

Preface

In this 21st century, a smart grid entails technology applications that will allow easier integration and higher penetration of renewable energy. It will be essential for accelerating the development and widespread usage in industrial & domestic domains, EVs, and potential use as energy storage. This book provides a comprehensive guide to researchers and students to design high-efficiency hybrid techniques and algorithms to overcome the issues in existing methods and technologies. It also includes energy measuring with the communication architectures. It aims to present prominent and emerging technologies that are available in the marketplace. It would emphasize and facilitate a greater understanding of all existing available research, i.e. theoretical, methodological, well-established, and validated empirical work, associated with the environmental energies.

The book focuses on simulation and practical development in multi-area realistic high-frequency power systems with stabilization of voltage and current by tuning the PID, fuzzy controllers, and artificial algorithms. In the off-grid, energy storage systems play a major role to maintain energy quality and stability. Solar and wall combustion, and generator installation, can also be integrated into hybrid ultracapacitor battery systems. The combination makes energy storage at both high energy and energy density, which extends the battery life. This hybrid storage has a wide demand in EVs and is conferred in the book. Multi-level inverters for high voltages applications in EVs and solar/wind energy harvesting are widely used to reduce the output harmonics of the system. DC converters are reliable and feasible for effective power conversion in grids. Types of converters, their working stability, and the short survey have been explained in detail. A study is conducted for effective harvesting of solar energy, the PV panel with Quantum-dot sensitized solar cells (QDSSCs). The grid-connected system has been embedded with IoT applications to monitor the operation in real time. For proper communication, wireless networks are also discussed.

The editors are grateful to the authors for their contribution to the book by illustrating the various hybrid methodologies in smart grid. We believe that this book has an important contribution to the community in assembling research work on developing on/off-grid systems. It is our sincere hope that many more will join us in this time-critical endeavor. Happy reading!

Mr. Seelam VSV Prabhu Deva Kumar, M.S.
Dr. Shyam Akashe, Ph.D.
Dr. Hee-Je Kim, Ph.D.
Dr. Chinmay Chakraborty, Ph.D.

MATLAB® is a registered trademark of The MathWorks, Inc. For product information,

please contact:
The MathWorks, Inc.
3 Apple Hill Drive
Natick, MA 01760-2098 USA
Tel: 508-647-7000
Fax: 508-647-7001
E-mail: info@mathworks.com
Web: www.mathworks.com

Editors

Seelam VSV Prabhu Deva Kumar is an assistant researcher (R&D) in JoongAng Control Co., Ltd. South Korea. He received Masters in Electrical Engineering from Pusan National University, South Korea, and Bachelors in Electronics and Communication Engineering, ITM University, M.P., India. He is also a certified chartered engineer (C.Eng) and member of IET (UK), IEI (India), and SESI (India). His research area focuses on DC–DC Converters and On/Off-Grid Inverters.

Shyam Akashe received Ph.D. in Electronics and Communication Engineering from Thapar University, Punjab, India. At present, he is professor of Department of Electronics and Communication Engineering in ITM University, Gwalior, India. He is a board member of *VLSI Circuits and Systems Letter*, *IEEE African Journal of Computing & ICTS*, IE, IETE, ISTE, ISOI, VSI. Biography published in Marquis's Who's Who in Engineering Field, USA awarded by Marquis, 2015. His research area is mainly in designing of low power CMOS circuits, modeling, FinFET-based memory design, implementation of control algorithms and digital design in FPGA boards, IoT applications, and low voltage DC–DC converters.

Hee-Je Kim received Ph.D. in Energy Conversion from Kyushu University, Japan. At present, he is professor in Department of Electrical Engineering, Pusan National University, South Korea. He is a Board Member of Journal [*Energies*], Korea Institute of Electrical Engineers and Renewable Energy. His research area is mainly in fabrication and commercialization of next-generation solar cells such as dye synthesized, perovskite and solar cells, improving efficiency of existing solar and wind hybrid systems using different techniques, high-energy and power density flexible super-capacitor for hybrid-energy storage systems, converters, MPPTs, PV inverters, and remote control by smartphone with novel algorithm for power conditioning systems.

Chinmay Chakraborty is an assistant professor (Sr.) in the Department of Electronics and Communication Engineering, BIT Mesra, India. His primary areas of research include wireless body area network, Internet of Medical Things, point-of-care diagnosis, smart city, green technology, m-health/e-health, and medical imaging. He has published more than 15 edited books and co-edited 10 books on smart IoMT, healthcare technology, and sensor data analytics with CRC Press, IET, Pan Stanford, and Springer. He has received the Young Research Excellence Award, Global Peer Review Award, Young Faculty Award, and Outstanding Researcher Award.

Contributors

Dr. Shyam Akashe was born in 22nd May 1976. He did M.Tech. in Electronics & Communication Engineering from Rajiv Gandhi Proudyogiki Vishwavidyalaya (RGPV), Bhopal in 2006 and Ph.D. from Thapar University, Patiala (previously Thapar Institute of Engineering and Technology), Punjab in 2013. Dr. Akashe has worked as assistant professor at the Institute of Technology & Management (ITM) Gwalior. Currently, he is working as professor and dean of R&D, Electronics & Communication Engineering Department at ITM University Gwalior. Dr. Akashe has authored/co-authored more than 200 research papers in peer-reviewed international/national journals and conferences. He has also published five patents, two copyrights, two books, and one chapter. His biography was published in Marquis's *Who's Who in Engineering Field, USA* awarded by Marquis, 2015. His areas of research are low power VLSI design, modeling, FinFET-based memory design, circuits for future VLSI technology, digital design, and FPGA implementation. He is life member of the Institution of Engineers (IE), life member of the Indian Society for Technical Education (IETE), life member of the Indian Society for Technical Education (ISTE) and life member of the Instrumentation Society of India (ISOI). He is also member of VLSI Society of India (VSI). Dr. Akashe is editorial board member of the *IEEE African Journal of Computing & ICTS* (IEEE Nigeria). He is also a reviewer for *IEEE Transaction on VLSI Systems*, *IEEE Transactions on Nanotechnology*, *International Journal of Electronics* (Taylor & Francis Group), *Microelectronics Journal* (Elsevier).

Dr. Suresh Alapati received his B.Tech. (Acharya Nagarjuna University, India) and M.Tech. (NIT, Calicut, India) degrees in Mechanical Engineering in 2004 and 2006, respectively. He obtained his Ph.D. (Dong-A University, Busan, Korea) degree in Mechanical Engineering in 2011. Dr. Suresh is presently working as an assistant professor in the Department of Mechatronics Engineering at the Kyungsung University, Busan, Korea. He worked as a postdoctoral fellow (2011–2012) in the Department of Mechanical Engineering at the Dong-A University, Busan, Korea. His research interests include computational fluid flow and heat transfer and simulation of micro/nano biological flows using the lattice Boltzmann and Brownian dynamics methods.

Ms. Sonalika Kishore Kumar Awasthi is pursuing her B.Tech. in Electronics and Communication Engineering with Specialization in Internet of Things and Sensors from Vellore Institute of Technology. Her areas of interest include the Internet of Things, computer architecture, embedded systems, microcontrollers, and fast computing.

Ms. M. Divya graduated from the Jawaharlal Nehru Technological University, Kakinada and completed her post-graduation in Vignan's University, Guntur. Presently she is working as an assistant professor in Vignan's Lara Institute of Technology & Science, Guntur and pursuing Ph.D. from Vignan's Foundation for

Science and Technology University, Guntur. She has six years of teaching experience. Her special fields of interest include power electronics and drives, and renewable energy resources.

Dr. Ikkurthi Kanaka Durga obtained her Master's and Ph.D. degrees in Electrical Engineering from Pusan National University, Busan, South Korea. He worked as a research assistant and senior researcher under Prof. Hee-Je Kim from March 2015 to February 2020. She obtained her bachelor of technology in information technology (IT) from Jawaharlal Nehru Technological University in Kakinada (2008–2012). She is very interested in global climate change and sustainable energy challenges and actively engaged myself in several related activities during my Master's and Ph.D. degrees. Currently, she is developing various efficient electrode materials that are used for the fabrication of energy storage and energy conversion devices. She has also published various research articles in SCI-based journals. She is an active reviewer for *Dalton Transactions*, *Electrochemica Acta*, and *Applied Surface Science*.

Dr. Geetha Reddy Evuri graduated in Electrical and Electronics Engineering from Jawaharlal Nehru Technological University, Hyderabad in 2007 and received her M.Tech. degree in Power Electronics and Drives from Vignan's University, Guntur in 2011. She completed her Ph.D. from Vignan's Foundation for Science and Technology University, Guntur in 2020. Presently she is working as an Associate professor in Teegala Krishna Reddy College of Engineering and Technology (TKRCET), Hyderabad. She has guided more than 15 B.Tech. and M.Tech. projects. She has also published more than ten papers in scopus/SCI indexed Journals. Her areas of interest include electric vehicles, renewable energy sources, control of power electronics and drives.

Mr. Srikanth Goud B received his B.Tech. degree in Electrical and Electronics Engineering and M.Tech. degree in Control Systems from JNT University, Hyderabad, in 2009 and 2012, respectively. He is currently working as assistant professor in Electrical and Electronics Engineering, Anurag College of Engineering, Hyderabad, India. He has submitted his Ph.D. thesis in Electrical and Electronics Engineering at Koneru Lakshmaiah Education Foundation. His research interests are integrated renewable energy systems, smart grids, power electronic converters, and their applications to power systems. He has published 25 research papers in various international journals and conferences which are indexed in *SCIE, Scopus, Web of Science, IEEE Xplore*. He has received the best paper award in NCRTPS-2014 organized by VKTS at Telangana, India.

Dr. Amit Grover obtained his Ph.D. from Guru Nanak Dev University Amritsar in 2021. He received his M.Tech. degree in Electronics and Communication Engineering from Punjab Technical University, Kapurthala, Punjab, India in 2008 and received his B.Tech. degree in Electronics and Communication Engineering from Punjab Technical University, Kapurthala, Punjab, India in 2001. Currently, he is working as an assistant professor in Shaheed Bhagat Singh State University, Ferozepur, Punjab,

India. His areas of interest include wireless sensor networks, signal processing, MIMO systems, wireless mobile communication, high-speed digital communications, free space optics (FSO), inter-satellite optical wireless systems (IsOWC) and VLSI design.

Mr. Meher Jangra, pursuing B.Tech. in Electronics and Communication Engineering with a specialization in Internet of Things and Sensors from Vellore Institute of Technology. His areas of interest are Internet of Things, computer architecture, embedded systems, microcontrollers, and fast computing.

Mr. CH. Naga Sai Kalyan received his M.Tech. degree from JNTUK, Kakinada, India. His research interests include the design of controllers for interconnected power system models using soft computing techniques and application of FACTS devices in power system operation and control. Currently he is working towards his doctoral degree.

Mr. Abhishek Kumar is currently working as an assistant professor in the Department of ECE of Swami Vivekanand Subharti University, Meerut, Uttar Pradesh. He had done his M.E. in ECE from BIT Mesra, Jharkhand. He completed his B.Tech. in ECE from Vinoba Bhave University, Hazaribag, Jharkhand. He has more than 30 international/national journals/conferences to his credit. His areas of interest include wireless communications, and wireless sensor networks.

Mr. Ankit Kumar (Member, IEEE) received his bachelor's degree in electronics and communication engineering from ITM University, India, in 2020. Currently, he is pursuing his integrated Ph.D. degree in electrical and electronics engineering from Hanyang University, South Korea. His research interest includes over-sampled delta-sigma converters.

Ms. Meet Kumari was born in Gurdaspur, Punjab, India on 11th, September 1990. She is pursuing her Ph.D. in Electronics and Communication Engineering from Punjabi University, Patiala, India. She completed her M.Tech. degree in Electronics and Communication Engineering from Guru Nanak Dev University, Regional Campus Jalandhar, Punjab, India in 2015 and received her B.Tech. degree in Electronics and Communication Engineering from Guru Nanak Dev University, Regional Campus Gurdaspur, Punjab, India in 2012. Currently, she is working as an assistant professor in Chandigarh University, Mohali, Punjab, India. She has published many papers/book chapters in reputed international journals/conferences. Her area of interests include optical communication, green communication, next-generation networks, and wireless sensor networks. She is a reviewer of many reputed international journals.

Mr. Seelam VSV Prabhu Deva Kumar, assistant researcher (R&D), JoongAng Control Co., Ltd., South Korea. He graduated from the Department of Electrical Engineering, Pusan National University, Busan, South Korea. His research works include inverters and DC converters for renewable and grid applications.

Dr. Jyoteesh Malhotra was born in Jalandhar, Punjab, India. He completed his B.Engg. *with Distinction* from P.R.M. Institute of Technology & Research, Amravati and M.Tech. with *University Gold Medal* from Guru Nanak Dev Engineering College, Ludhiana. He received his Ph.D. from Panjab Technical University in recognition to his contribution in the field of Wireless Communication & Networks. From 1994 to 2007, he was employed with DAVCMC, New Delhi and Panjab University, Chandigarh. He joined Guru Nanak Dev University Regional Campus at Jalandhar in 2007 where he is currently professor and dean of the Faculty of Engineering & Technology. His research interests are in the broad area of Internet of Things, Machine Learning application in Pervasive Wireless Communication & Networks, and Optical Communication. Dr. Malhotra has published and presented more than 200 technical papers in scientific journals and international conferences and authored 02 books. He is a life member of Indian Society for Technical Education (I.S.T.E.) and Editorial Board of many international journals of repute.

Dr. Vishwas Mishra is currently working as an assistant professor in the Department of ECE of Swami Vivekanand Subharti University, Meerut, Uttar Pradesh.

Sasikumar Periyasamy, professor, VIT University, was awarded a Ph.D. in QOS Enhancement for Distributed Clustering Techniques in Wireless Sensor Networks, VIT University, a M.E. in Applied Electronics from Government College of Technology (Autonomous Institution), and B.E. in Electrical and Electronics from Anna University. He has expertise in the areas of embedded systems, wireless sensor networks, communication engineering, Internet of Things and security.

Dr. Kummara Venkata Guru Raghavendra received his bachelor's degree in Electrical and Electronics Engineering from JNTU, Anantapur, India. He completed his Master's in Automation Engineering from University of Bologna. In August 2020, he received his Ph.D. from Pusan National University, with a major in Electrical Engineering. He worked as a post-doctoral researcher in Yeungnam University. He is a potential reviewer for *Journal of Energy Storage, Materials Letters*, and *Ceramics International*. He is member of many reputed research organizations such as IEEE, International Society of Automation, Franklin UK, the IRED. Currently he is a sustainable energies research and project consultant and visiting post-doctoral researcher in Renewable Engineering at American University of Ras Al Khaimah University, RAKRIC, UAE.

Dr. G. Sambasiva Rao received his Ph.D. from JNTUH, Hyderabad, India in 2014. His research interests include controlling techniques for dual inverter fed open end winding induction motor, FACTs Controllers, and power quality improvement. Currently he is working as professor in EEE Department of RVR&JC College of Engineering, India.

G. Srinivasa Rao

Prof. Ramesh Chandra Rath is an eminent professor & academician of an international repute in the field of Management Education (Mkt. & HR) and Clinical

Psychology. He has served 27 years of Service in various government college and universities of Odisha and abroad. He obtained his Post Graduate Degree in Clinical Psychology from SCB Medical College, Cuttack, which is affiliated with Utkal University, Bhubaneswar, and his MA in Psychology from Sambalpur University. He completed his MBA degree from Delhi University in 1996, his Ph.D. degree from BIT Mesra, Ranchi in 2000 on the area of research "Green Marketing & Supply Chain Management" (Marketing Management), and his Postdoctorate Degree (D.Litt.) from Patna University, Patna, Bihar in 2003 in the area of "Advanced Physiology and Criminology". Currently, he is working as a professor-cum-dean (Research & Development), and head of the Department of MBA at GGSESTC, Bokaro, Jharkhand, India. He has guided 17 Ph.D. research scholars of various universities in the areas of green marketing & supply chain management, consumer behavior, and organizational behavior, production and operational management. There are 32 international journals and 38 national journals with two books of publication in his credit. Prof. Rath is an acting member of editorial boards of many reputed national and international journals including *IEEE, IEOM, IJCEM, IJEBM, IJSR, IJSER, Science PG, Springer*, and *Elsevier* for reviewing papers and conference proceedings. He was also invited as an external examiner for Ph.D. thesis evaluation, and grand VIVA-VOCE (Ph.D.) for many universities of national and international repute. He has 65 Ph.D. theses examined as an external examiner. He has also been invited regularly as resource person, keynote speaker, and session chair for various international and national conferences, seminars, and workshops.

Dr. B. Nagi Reddy received his B.Tech. degree in Electrical and Electronics Engineering, in 2010 and M.Tech. degree in Power Electronics, in January 2013 from Jawaharlal Nehru Technological University, Hyderabad. He completed his full-Time Ph.D. from Koneru Lakshmaiah Education Foundation (KLEF – Formerly K L University) in 2020. He has worked as assistant professor and HOD in Brilliant Group of Technical Institutions at Hyderabad from January 2013 to January 2016. During this tenure he guided more than ten B.Tech. & M.Tech. Projects. He has assisted a book called "Electrical Machines," written by Dr. M. Ramamurthy & Dr. O. Chandra Sekhar, which was published in 2017. He is the co-author of a monograph on "Reactive Power" published in 2020. He has also published more than 15 Technical papers in the SCI & Scopus indexing journals.

He is currently working as associate professor in the Electrical and Electronics Engineering Department in Vignana Bharathi Institute of Technology (VBIT), Hyderabad. His area of interests are design & control of power converters & drives, electric vehicals, and renewable energy sources.

Ch Rami Reddy

Dr. K. Srinivasa Reddy received his Ph.D. from OPJS University, Churu, Rajasthan, M.Tech. degree in Embedded Systems and B.Tech. degree in Electronics and Communication Engineering from Jawaharlal Nehru Technological University, Hyderabad. Currently, he is working as an associate professor in the Department of Electronics and Communication Engineering, Teegala Krishna Reddy Engineering College, Hyderabad. He has guided more than 25 B.Tech. and 15 M.Tech. projects.

He has also published more than 08 papers in Scopus/SCI indexed journals and four textbooks. His areas of interest are communication systems, cellular and mobile communication, wireless communications & networks, and applications fields of embedded systems.

Dr. Reecha Sharma obtained her Bachelor's degree in Electronics and Instrumentation Engineering from MMEC Mullana, Haryana, India in 2003 and Master's degree in Electronics Instrumentation and Control Engineering from T.I.E.T., Patiala, Punjab, India in 2005. She obtained her Ph.D. degree from Punjabi University, Patiala, in 2016. Presently, she is working as assistant professor in Punjabi University, Patiala. Her current interests are digital image processing, digital signal processing, feature extraction, face recognition, filter designing, opto-imaging, control engineering, fuzzy logics, artificial neural networks, and optimization techniques. She has over 178 research papers published/presented in international/national journals/conferences to her credit. She is also a reviewer of many reputed international journals.

Dr. Anu Sheetal was born in Bahadurgarh, Haryana, India, on 18 September 1972. She obtained her Bachelor's degree in Electronics Engineering with distinction from the Department of Electronics & Communication Engineering, Punjabi University, Patiala, India in 1994 and Master's degree in Electronics Engineering from Punjab Technical University, Jalandhar, India in 2003. She obtained her Ph.D. degree from Punjab Technical University, Jalandhar, in 2012. She has worked as a design engineer at Gilard Electronics Private Limited, Mohali, from 1994 to 1997. She then joined AIET, Faridkot as a lecturer and became head of the Department in 1999. In 2004, she joined Guru Nanak Dev University, Regional campus, Gurdaspur, Punjab, India in the Department of Electronics and Communication Engineering as a lecturer and became assistant professor in the Department of Electronics and Communication Engineering. Presently, she is working as senior assistant professor and head of the Department in the same institute. Her current interests are optical communication systems, soliton transmission and DWDM Networks, WDM-PON, and RoF. She has over 50 research papers published/presented in international/national journals/conferences to her credit. She is a life member of The Institution of Electronics and Telecommunication Engineers (IETE), New Delhi (India), Indian Society of Technical Education (ISTE), Optical Society of India, Kolkata (OSI), and International Association of Engineers (IAENG). She is acting as a member of editorial board of *The International Journal of VLSI & Signal Processing Applications (IJVSPA)*, *International Journal of Research IJRICE*, and *ISP Journal of Electronics Engineering*. She is acting as technical reviewer for many international journals such as *Journal of SPIE – Optical Engineering* and *OPTIK*.

Mr. Mehtab Singh received his Bachelor of Engineering in Electronics and Communication Engineering from Thapar University, Patiala and Master of Technology in Electronics and Communication Engineering with specialization in Communication Systems from Guru Nanak Dev University, Amritsar. He is currently pursuing his Ph.D. from Guru Nanak Dev University, Regional Campus,

Jalandhar. His areas of interest include free space optics and inter satellite optical wireless communication links.

Dr. Amita Soni is currently working as an assistant professor in the Department of Computational Sciences of Christ University (Deemed to be University), Delhi NCR. She has completed her Ph.D. and M.Sc. in Mathematics from NIT Rourkela, Odisha. She received a B.Sc. in Mathematics (Hons.) from St. Xavier College, Ranchi, Jharkhand. She is gold medalist in her M.Sc. Her areas of interest include mathematical modeling, elliptic PDEs, wireless sensor networks, and artificial intelligence (AI).

Dr. Gaurav Srivastava works at Poornima College of Engineering. He is currently working as an assistant professor in the Department of Electrical Engineering and has been associated with the institute since July 2014. He has vast experience spanning around 11 years both in industry and in academics. He has completed his B.Tech. (2010) in Electrical Engineering, and M.Tech. (2014) in Power Systems from Nirma University, Ahmedabad. His Ph.D. (2020) is in renewable energy and solar power. He is a lifetime member of IEANG, ISTE, ISRS, and has published over 30 research papers in reputed conferences and journals (*Scopus*, *SCI*, *UGC Care*). He also bagged five Indian patents and one Australian patent on his name. He has authored a book entitled "Emerging research trends in PV Technologies" with USA ISBN and number of chapters were published in various reputed books.

Dr. Sanjeev Kumar Srivastava is working as an associate professor and Head of the Department of Electronics and Telecommunication Engineering at Pillai College of Engineering, Mumbai since 2008. On the professional front, he has more than 24 years teaching experience as an approved faculty of Mumbai University and has served in all capacities in various reputed Engineering colleges affiliated to Mumbai University. In 2018, he was awarded a doctorate degree from Nagpur University for his Ph.D. thesis work, "Study and Analysis of Ad-hoc Wireless Network System to Maximize the Performance of Throughput Capacity". He has also published more than 25 research papers in reputed national and international conferences and journals. Some of his research papers have been indexed in IEEE and Springer journals. He got the best research paper award in the IEEE international conference conducted by Jamia Millia Islamia University, New Delhi. He has also been invited as judge in Technical Paper Presentation and Project Exhibition in several national level events.

He is an approved and recognized teacher in various universities to guide undergraduate and postgraduate students and will be eligible soon to also guide Ph.D. students. He has also been appointed the chairman of Vigilance Squad Committee in smooth conduction of examinations held by Mumbai University. He is also the paper setter and evaluator of various subjects related to electronics and telecommunication engineering in various universities. He is also appointed as research paper reviewer on the editorial board of various reputed IEEE international conferences and journals related to the communication engineering field.

Dr. Srinivasa Rao Sunkara is an assistant professor of Mechanical and Mechatronics Engineering at KyungSung University in Busan, South Korea. He obtained his undergraduate degree in Electrical and Communication Engineering (2007–2011) from Jawaharlal Nehru Technological University in Kakinada, India. He obtained his Master's and Ph.D. in Electrical Engineering from Pusan National University in 2015 and 2018, respectively. He has served as a research scientist at Pusan National University from 2012 to March 2018, respectively, where he was involved in the development of various nanomaterials which can be used in Energy Storage and Conversion applications. Since March 2018, he has worked as an assistant professor in the School of Mechanical and Mechatronics Engineering., KyungSung University, South Korea. Currently, he is developing various electrode materials for supercapacitors, solar cell devices using novel nanomaterials. He is also an active reviewer for many SCI-based journals such as *New Journal of Chemistry*, *Electrochimica Acta*, *Dalton Transactions*, *RSC Advances*, *Applied Surface Science*, *Journal of Energy Storage*, *Langmuir*, *Heliyon*, and *ACS Applied Materials & Interfaces*. He is also an Guest editor for various journals such as *Energies (MDPI)* and *Frontiers in Materials*.

Mr. Shobhit Tyagi is currently working as an assistant professor in the Department of ECE of Swami Vivekanand Subharti University, Meerut, Uttar Pradesh. He had done his M.Tech. in ECE from RGTU, Bhopal. He received a B.Tech. in ECE from AKTU, Uttar Pradesh. He has more than ten international/national journals/conferences to his credit. His areas of interest include wireless communications and wireless sensor Networks.

Prof. Kisoo Yoo did his M.S. in Mechanical Engineering from Korea Advanced Institute of Science and Technology, March 2006–August 2008, and worked later as a research assistant at Korea Institute of Energy Research in Energy Efficiency Department to Develop a APU system using a solid oxide fuel cell. Modeling a water flooding problem on Direct Methanol Fuel cell. He also worked on Developing a centrifugal pump for submarine at Korea Institute of Machinery and Materials from the Department of Mechanical Engineering. He completed his Ph.D. from Microscale Thermo-Fluid Lab Washington State University, USA. Prof. Kisoo is an assistant professor in the School of Mechanical Engineering, Yeungnam University, South Korea. He has published in various national and international journals with higher citations.

1 Performance Comparison of Various FACTS Devices in Combined LFC and AVR of Multi-Area System for Simultaneous Frequency and Voltage Stabilization

CH. Naga Sai Kalyan
Acharya Nagarjuna University

G. Sambasiva Rao
RVR&JC College of Engineering

CONTENTS

DOI: 10.1201/9781003143802-1

1

1.1   INTRODUCTION

Modern power systems are becoming more complex due to the interconnection of several diverse source generation units to meet the varying load demand. Maintaining the stability of these complex systems is not an easy task. Moreover, system stability depends on regulating the frequency by diminishing real power gap between generation and demand. This action can be performed by the load frequency controller (LFC). However, reliable and efficient functioning of power systems also relies on keeping system terminal voltage within specified limits. This task will be governed through automatic voltage regulator (AVR) by altering generator field current. So, system stability can be restrained by getting control over system frequency and voltage at terminal end. Thus, these two quantities must be addressed in a combined manner to attain system stability more economically and efficiently.

Researchers had contributed a lot of investigations on LFC and AVR separately. The concept of LFC was first proposed by Fosha and Elgerd (1970) for thermal systems; after that many researchers have started focusing on frequency stabilization of interconnected power systems. Guha et al. (2016) investigated the thermal system taking into account the without and with constraints of non-linearity. Arya (2019) had performed the investigations on hydrothermal unit of multi-area system. Gas power systems had also been considered with hydrothermal unit having structure of reheat turbines for stability analysis of two-area system by Shankar et al. (2016). Renewable power conversion sources like harnessing power from solar and wind integration with the classical generation units of hydrothermal system had been considered for investigation in Bheem and Deepak (2019), but the physical parameters of generation rate constraints (GRCs) are not taken into account. Although Fathy and Kaseem (2018) considered the parameters of non-linearity along renewable integration to the existing conventional interconnected system, their investigations are limited to only frequency regulation and the coupling of AVR is not considered.

The literature on combined LFC and AVR analysis is available and bounded to a conventional scenario. Rakshani et al. (2009) carried out the combined analysis on thermal power plants but confined to a single area only. Later on, it was extended to a two-area system by Chandrakala and Balamurugan (2016) and renewable integration was not examined as they are gaining momentum. Rajbongshi and Saikia (2017, 2018) carried out the combined analysis on multi-area system with integration of solar thermal plant, but wind turbine integration was not considered. Hence, this promotes the author to investigate the combined analysis of the multi-area system comprising photovoltaic and wind turbines.

However, designing of controller for LFC and AVR loops plays a critical role in maintaining the stability of the system and is also the most recent trend of research. Different classical controllers like PI (Magid and Abido 2003)/PID (Gozde et al. 2012), fractional order (FO) FOPI/FOPID (Tasnin and Saikia 2018), intelligent fuzzy (Arya 2019) and artificial neural network (ANN) (Fathy and Kaseem 2018) controllers are available in the literature. In this present work, classical PID is used for both LFC and AVR loops as the secondary regulator because of simplicity in design, efficiency in operation, and because it requires less number of knobs for controlling purpose when compared to FO controllers and also requires less quantity of data compared to implementation of fuzzy controllers. Moreover, proper selection of searching algorithm for controller optimization brings efficacy and robustness. Various optimization algorithms available in the literature are differential evolution (DE), particle swarm optimization (PSO) (Magid and Abido 2003), artificial bee colony (ABC) (Gozde et al. 2012), ant lion optimizer (ALO) (Fathy and Kaseem 2018), bacteria foraging optimization (BFOA) (Ali and Elazim 2013), grey wolf optimization (GWO) (Guha et al. 2016) etc. But most of these algorithms had been trapped into local minima, premature convergence and not possessing the tendency of keeping average equilibrium between exploitation and exploration. So, a new DE-AEFA algorithm is presented in this chapter to solve the combined model of the realistic power system.

For large load disturbances, the secondary regulator is not potential enough to drag the system into the stability region. So, the system needs an additional control strategy of FACTS and energy storage devices (ESDs). This chapter addresses the performance analysis of several FACTS devices like Thyristor controlled series capacitor (TCSC), static synchronous series compensator (SSSC), Thyristor controlled phase shifter (TCPS) and Interline power flow controller (IPFC) and is not available in the literature so far on combined effect. Finally, the coordinated control mechanism of best FACTS device obtained from above with RFBs is implemented to further mitigate the system deviations.

The major contributions of this work are as follows.

a. LFC and AVR combined model of realistic two-area power system is designed for investigation.
b. PID controller is optimized with DE, AEFA and DE-AEFA algorithms; best suitable controller is demonstrated by comparing the dynamic performance.
c. Performance of the DE-AEFA presented algorithm is validated and tested on the conventional power system model available in the literature.
d. The effect of AVR coupling on LFC is revealed and justified.
e. Performance comparison of several FACTS devices on the combined model is investigated and the superior performance of IPFC over others is exposed.
f. Control of IPFC-RFBs coordinated mechanism is implemented to damp out the system deviations further.

1.2 CONSIDERED POWER SYSTEM MODELS

1.2.1 POWER SYSTEM UNDER INVESTIGATION

Equal generation capacities of two-area thermal power plant with reheat turbine structure named as test system-1 is considered for investigation to validate the efficacy of presented DE-AEFA optimized PID controller. The parametric gains and

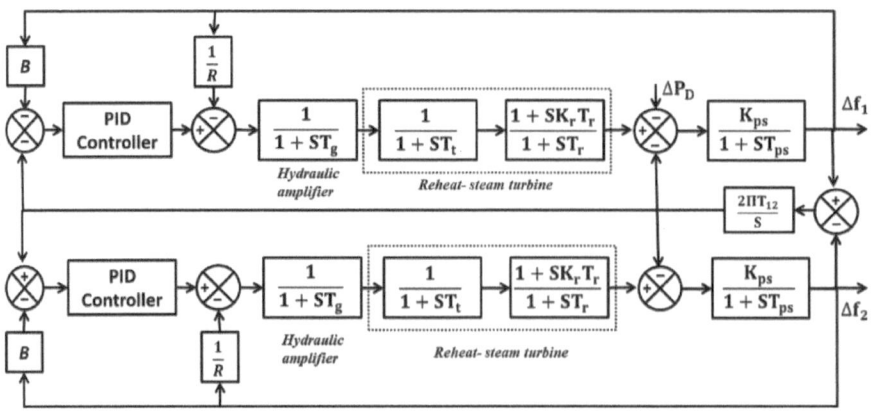

FIGURE 1.1   Model of two-area thermal system with reheat turbine structure (test system-1).

time constants of test system-1 are directly considered from the study by Ali and Elazim (2013). Another test system of combined LFC and AVR model named test system-2 considered for investigation consists of two unequal areas having a 2:1 ratio of generation capacities. Control area-1 consists of reheat thermal unit, hydro and gas generations. Diesel, wind and solar photovoltaic generation are incorporated in control area-2. The test system-1 under investigation is depicted in Figure 1.1 and test system-2 is depicted in Figure 1.2. For hydro and thermal units non-linearity parameters of GRC is considered to make power system more realistic. The parameters of test system-2 are considered from the work of Shankar et al. (2016) and Rajbongshi and Saikia (2017). The coupling of AVR loop with LFC is carried out through cross coupling coefficients depicted in Figure 1.3.

1.2.2   MODELING OF COMBINED LFC AND AVR LOOPS

The main role of combined LFC and AVR is to maintain the interconnected power system stable by maintaining the system frequency within the limits and generator terminal voltage at a specified value under load variations. The interaction between LFC and AVR exists with the cross coupling coefficients K_1, K_2, K_3 and K_4. In the combined model, the frequency and real power control is taken by the LFC loop, whereas the terminal voltage is maintained by the AVR loop. The generator terminal voltage (V) is continuously monitored by the sensor unit and compared with the specified voltage. Difference in voltage is taken as error and is used to change the field excitation after amplification. These controlling measures in AVR loop affect the magnitude of EMF building up in the generator armature terminals $\left(E'\right)$ which successively influence real power generation given by

$$P_e = \frac{|V||E'|'}{X_S}\sin(\delta) \tag{1.1}$$

Hence, the AVR loop has a significant and simultaneous impact on LFC loop.

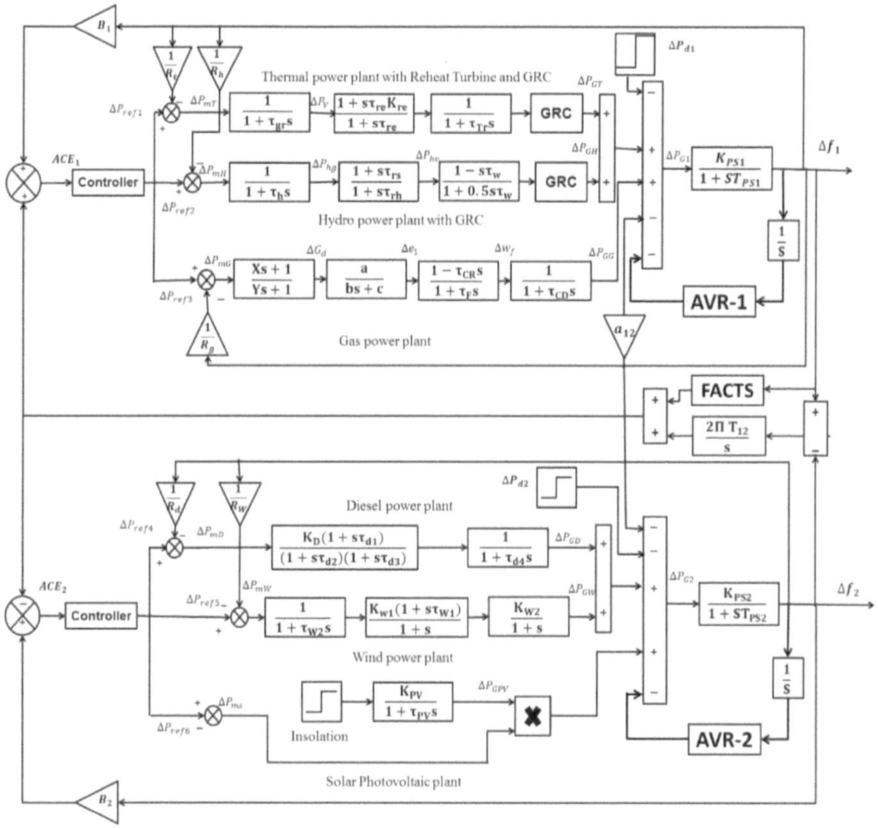

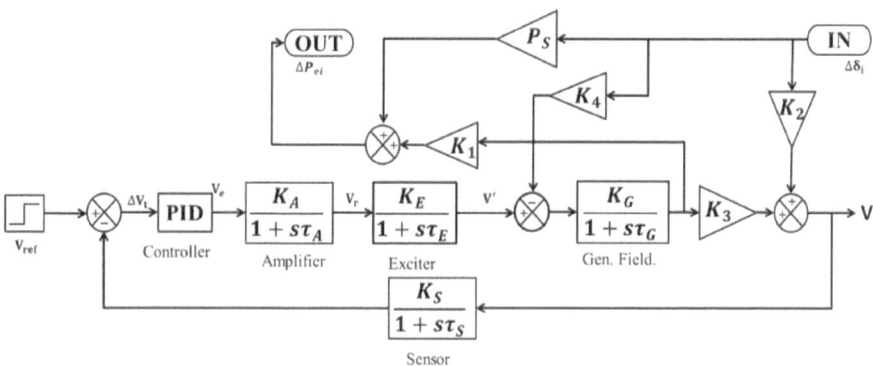

FIGURE 1.2 Combined LFC and AVR model of multi-area hybrid system (test system-2).

FIGURE 1.3 AVR with cross-coupling coefficients.

During small load disturbance, the frequency deviation can be controlled by changing the generator rotor angle and hence the change in real power generation is given by

$$\Delta P_e = P_S \Delta\delta + K_1 E' \qquad (1.2)$$

where P_S is synchronizing power coefficient, K_1 is change in real power with change in E' at constant δ. The terminal voltage (V) is composed of two components. One is q-axis component (V_q) which changes with E' and other is d − axis component (V_d) which changes with rotor angle δ.

Then, change in terminal voltage is expressed as

$$\Delta V = K_2 \Delta\delta + K_3 \Delta E' \qquad (1.3)$$

where K_3 is change in V for a change in E' at constant load angle, K_2 is change in V for a change in δ at constant E'. The factors that affect the voltage induced in the generator are given by

$$E' = \frac{K_G}{1 + S\tau_G}(V' - K_4 \Delta\delta) \qquad (1.4)$$

where K_4 is variation in E' for a small change in δ, and K_G and τ_G are generator field gain and time constants, respectively.

1.3 MODELING OF FACTS DEVICES

FACTS devices are recent revolutionary apparatus emerging from the power electronics family. These devices are playing a crucial role in controlling the flow of power in AC lines. The power oscillations can be effectively damped by placing FACTS devices with the tie-line. Hence, system stability can be effectively achieved. By regulating tie-line power oscillations system frequency and terminal voltages will be maintained at specified limits. Moreover, these devices also offer enhancement in system capacity and reliability. The mathematical modeling of these FACTS devices are considered from Lal and Barisal (2017) and are applied to combined LFC and AVR, which is discussed in detail in this section.

1.3.1 Static Synchronous Series Compensator (SSSC)

SSSC is configured by connecting a transformer in series with a transmission line in which controlling of power exchange is essential. SSSC has a very wide range of operating characteristics as it provides both series capacitive or inductive compensations. This can be achieved by supplying a balanced voltage with quadrature component of line current at fundamental frequency. With SSSC, the power exchange between control areas can be modeled as

$$\Delta P_{\text{tie12}}(S) = \Delta P_{\text{tie12}}^0(S) + \Delta P_{\text{SSSC}}(S) \qquad (1.5)$$

$$\Delta P_{\text{tie}12}^0(S) = \frac{2\pi T_{12}}{S}\left(\Delta f_1(S) - \Delta f_2(S)\right) \tag{1.6}$$

$$\Delta P_{\text{SSSC}}(S) = \frac{K_{\text{SSSC}}}{1 + ST_{\text{SSSC}}}\left(\frac{1 + ST_1}{1 + ST_2}\right)\left(\frac{1 + ST_3}{1 + ST_4}\right)\Delta\text{Error}(S) \tag{1.7}$$

The detailed structure of SSSC as damping controller and the single-line diagram of the SSSC is shown in Figures 1.4 and 1.5, respectively. The time lag between input and output signals can be compensated by T_1–T_4 time constants.

1.3.2 THYRISTOR CONTROLLED PHASE SHIFTER (TCPS)

This is a series connected device which can be used to control power flow by influencing respective bus voltage angles. In general, it is located nearer to area-1 and line resistance is neglected as the X/R ratio of transmission line is high. The structure of this device is shown in Figure 1.6. The power exchange between control areas without considering TCPS can be expressed as

$$\Delta P_{\text{tie}12}^0(S) = \frac{2\pi T_{12}}{S}\left(\Delta f_1(S) - \Delta f_2(S)\right) \tag{1.8}$$

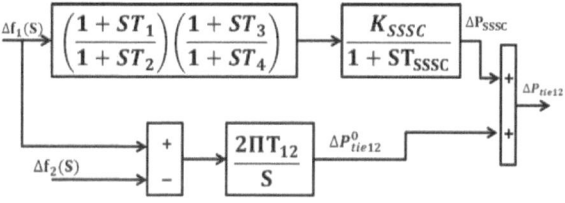

FIGURE 1.4 SSSC as a damping controller.

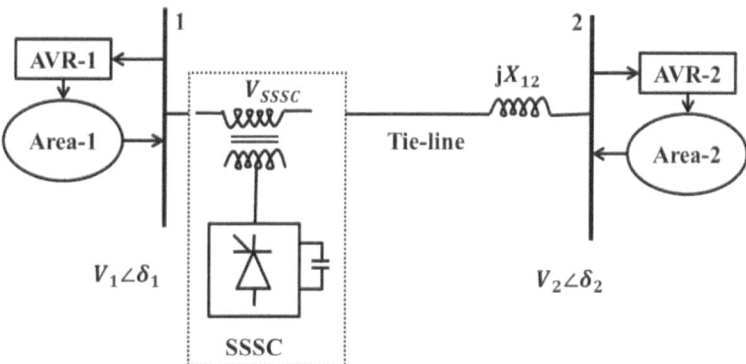

FIGURE 1.5 Single-line representation of combined system with SSSC in series with tie-line.

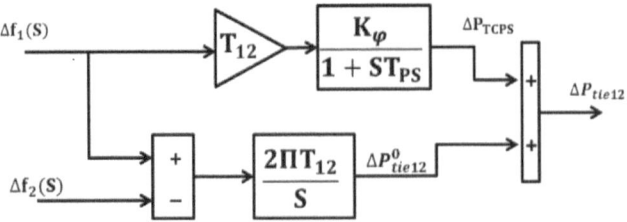

FIGURE 1.6   TCPS as damping controller.

The detailed analysis of this device is presented in Shankar et al. (2016). The shift in phase angle by considering this device can be expressed as

$$\Delta\varphi(S) = \frac{K_\varphi}{1 + ST_{PS}} \Delta\text{Error}(S) \tag{1.9}$$

where T_{PS} and K_ϕ are time and gain constants of TCPS.

The incremental power flow between the control areas with TCPS can be expressed as

$$\Delta P_{\text{tie}12}(S) = \frac{2\pi T_{12}}{S}\big(\Delta f_1(S) - \Delta f_2(S)\big) + T_{12}\frac{K_\varphi}{1 + ST_{PS}} \Delta\text{Error}(S) \tag{1.10}$$

Here, the frequency deviation is considered as input error signal to TCPS; then the above equation becomes

$$\Delta P_{\text{tie}12}(S) = \frac{2\pi T_{12}}{S}\big(\Delta f_1(S) - \Delta f_2(S)\big) + T_{12}\frac{K_\varphi}{1 + ST_{PS}} \Delta f_1(S) \tag{1.11}$$

$$\text{now } \Delta P_{\text{tie}12}(S) = \Delta P_{\text{tie}12}^0(S) + \Delta P_{\text{TCPS}}(S) \tag{1.12}$$

$$\text{Where }\ \Delta P_{\text{TCPS}}(S) = T_{12}\frac{K_\varphi}{1 + ST_{PS}} \Delta f_1(S) \tag{1.13}$$

Single-line diagram of the test system under investigation with TCPS is shown in Figure 1.7.

1.3.3   THYRISTOR CONTROLLED SERIES CAPACITOR (TCSC)

TCSC is a series device with the combination of shunt capacitor and Thyristor controlled reactor. It controls the dynamic power flow in a transmission line and improves the stability of the system and power quality. The total effective reactance of the line is controlled using TCSC by varying reactive power injected by the capacitor. The system with TCSC as damping controller is shown in Figure 1.8.

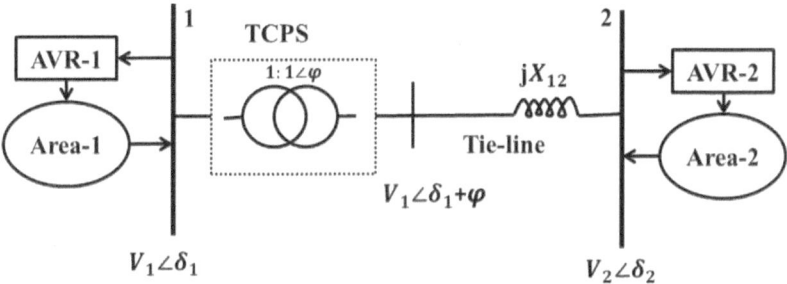

FIGURE 1.7 Single-line representation of combined system with TCPS.

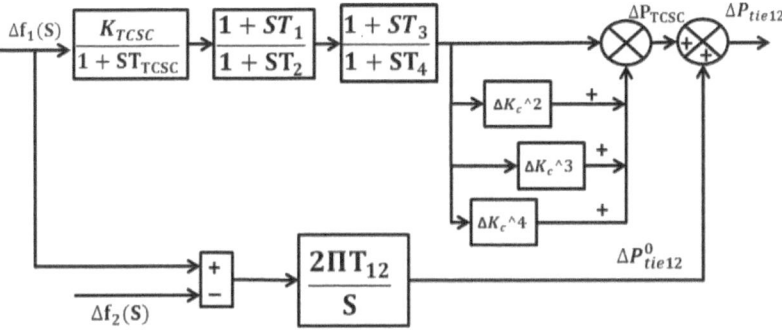

FIGURE 1.8 TCSC as a damping controller.

Incremental power between areas with TCSC can be expressed as

$$\Delta P_{\text{tie}12}(S) = \Delta P_{\text{tie}12}^{0}(S) + \Delta P_{\text{TCSC}}(S) \tag{1.14}$$

where $\Delta P_{\text{tie}12}^{0}$ represents the incremental power flow through tie-line without and ΔP_{TCSC} indicates the effect of TCSC device in power flow between areas.

$$\Delta P_{\text{TCSC}} = \Delta K_C + \Delta K_C^2 + \Delta K_C^3 + \cdots \tag{1.15}$$

From Equation 1.15, the incremental power flow through tie-line power flow can be controlled by adjusting ΔK_C and this can be expressed as

$$\Delta K_C(S) = \frac{K_{\text{TCSC}}}{1 + ST_{\text{TCSC}}} \left(\frac{1 + ST_1}{1 + ST_2}\right)\left(\frac{1 + ST_3}{1 + ST_4}\right)\Delta \text{Error}(S) \tag{1.16}$$

The combined model with TCSC is represented as single-line diagram and is depicted in Figure 1.9; deviation in area-1 frequency is considered as control signal. In Equation 1.15, only three terms are approximated for SIMULINK design to avoid unnecessary simulation burden.

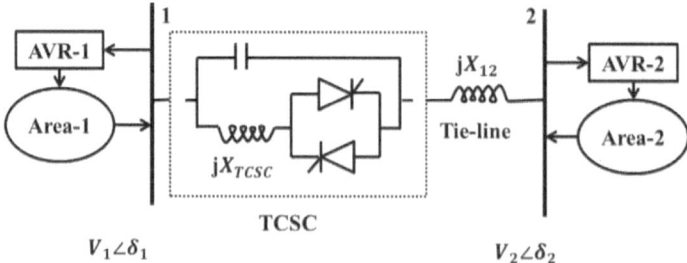

FIGURE 1.9 Single-line representation of combined system with TCSC.

1.3.4 INTERLINE POWER FLOW CONTROLLER (IPFC)

IPFC is usually facilitated to multiple transmission lines compensation at substations. All Thyristor and SSSC based controllers enhance the real power transfer capability by varying the line reactance, and they are pathetic in reactive power regulation. But IPFC has the capability of providing highly efficient scheme for power transfer management in transmission lines. The schematic model of IPFC as damping controller and single-line representation is shown in Figures 1.10 and 1.11, respectively

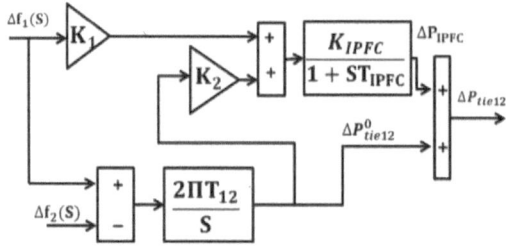

FIGURE 1.10 IPFC as damping controller.

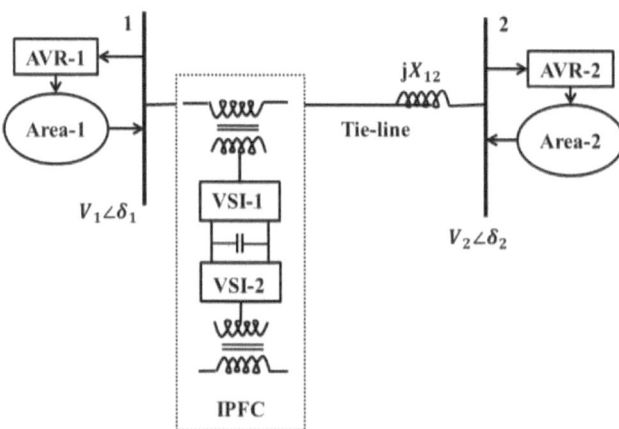

FIGURE 1.11 Single-line representation of combined system with IPFC.

The incremental power through tie-lines in an interconnected system can be expressed as

$$\Delta P_{\text{tie}12}(S) = \Delta P_{\text{tie}12}^{0}(S) + \Delta P_{\text{IPFC}}(S) \tag{1.17}$$

The impact of IPFC on power flow through tie-line can be expressed as

$$\Delta P_{\text{IPFC}}(S) = \left(\frac{K_{\text{IPFC}}}{1 + ST_{\text{IPFC}}} \right) \left(K_1 \Delta f_1(S) + K_2 \Delta P_{\text{tie}12}^{0}(S) \right) \tag{1.18}$$

1.4 REDOX FLOW BATTERIES (RFBS)

RFBs are the most eminent energy storage devices using static storage technology. The charging and discharging process in RFBs are done through the chemical process of oxidation and reduction reactions. RFBs possess a reactor tank having dual compartments. They are segregated with a proton membrane and facilitate to link up with a reservoir tank via pump, helping to circulate the electrolytic solution. The reactor in RFBs is filled with an amalgamation of sulfuric acid and vanadium pentoxide as electrolyte. The energy is maintained continuously by pumping fresh electrolyte into the reactor tank whenever necessary. The stored energy from the battery will be instantly delivered to the system under peak demands. The rectifier and inverter action will be taken care by the dual converter. The significant features of RFBs include quick response capability, ease of operation, less losses, low maintenance due to static technology and high life expectancy designating the superiority of RFBs over other energy storage devices. The transfer function of RFBs is considered from Tasnin and Saikia (2018).

$$G_{\text{RFB}} = \frac{K_{\text{RFB}}}{1 + ST_{\text{RFB}}} \tag{1.19}$$

1.5 OBJECTIVE FUNCTION FOR CONTROLLER DESIGN

The integral square error (ISE) performance index is considered to elevate the performance of proposed DE-AEFA algorithm. The objective function to be minimized in dynamic analysis of combined LFC and AVR is ISE. The parameters of PID controller are tuned to minimize this error using proposed algorithm subjected to satisfying parameter constraints.

$$J = \int_{0}^{T_{\text{sim}}} \left[\Delta f_1^2 + \Delta f_2^2 + \Delta P_{\text{tie}1,2}^2 + \Delta V_1^2 + \Delta V_2^2 \right] dt \tag{1.20}$$

The parameter constraints considered in optimization process are as follows:

$$K_{P\min} \leq K_P \leq K_{p\max}, K_{i\min} \leq K_i \leq K_{i\max}, K_{d\min} \leq K_d \leq K_{d\max},$$

$$K_{\text{SSSCmin}} \leq K_{\text{SSSC}} \leq K_{\text{SSSCmax}}, K_{\text{TCSCmin}} \leq K_{\text{TCSC}} \leq K_{\text{TCSCmax}}, K_{\text{IPFCmin}} \leq K_{\text{IPFC}} \leq K_{\text{IPFCmax}},$$

$$T_{\text{SSSCmin}} \leq T_{\text{SSSC}} \leq T_{\text{SSSCmax}}, T_{\text{IPFCmin}} \leq T_{\text{IPFC}} \leq T_{\text{IPFCmax}}, T_{\text{TCSCmin}} \leq T_{\text{TCSC}} \leq T_{\text{TCSCmax}},$$

$$T_{1\text{min}} \leq T_1 \leq T_{1\text{max}}, T_{2\text{min}} \leq T_2 \leq T_{2\text{max}}, T_{3\text{min}} \leq T_3 \leq T_{3\text{max}}, T_{4\text{min}} \leq T_4 \leq T_{4\text{max}} \quad (1.21)$$

The maximum and minimum limit of the PID controller gains are [0–5] and those of FACTS devices are [0–2]; the time constants of FACTS devices and time lag compensation coefficients are considered to be between [0.001 and 1].

1.6 DIFFERENTIAL EVOLUTION-ARTIFICIAL ELECTRIC FIELD ALGORITHM (DE-AEFA)

Differential evolution (DE) was proposed by Storn and Price (1997) and falls under the category of evolutionary algorithms anchored on stochastic search methods. DE initializes population individually and each population evaluates the fitness according to the problem to be solved. In each iteration, it generates a new population based on the selection process; the population having a better fitness value may participate in the new population for the next iteration. This procedure will continue until the stopping criteria is met or maximum iterations are reached. However, every optimization algorithm has its own benefits and weakness. The major strengths of DE are population upgradation and efficiency in local searching process. Moreover, instability in convergence and a tendency to easily drop into local best solution are its weaknesses. Artificial electric field algorithm (AEFA) was proposed by Anita and Yadav (2019) inspired from Coulomb's law of electrostatic force. Analyzing the application of AEFA algorithm to LFC problem it came to be known that it has the quality of quick convergence and capability of memorization. In addition to above, it shows superior performance at initial iterations but indicates stagnancy at large iterations.

The problem considered in this work is more complex and non-linear, so optimizing such problems definitely needs a supreme algorithm with efficient searching capability, quick convergence and less computational burden. Keeping these characteristics in mind a new combined DE and AEFA algorithm is proposed in this chapter. Blending of DE and AEFA algorithms results in DE-AEFA formation which possesses the benefits of both the algorithms and yields best results in quick time by keeping the average equilibrium among exploitation and exploration.

The implementation methodology of combined DE-AEFA algorithm is

Step 1: Initialize initial parameters, number of populations in DE (N_{DE}); number of populations in AEFA (N_{AEFA}); total number of populations in DE-AEFA is ($N_{\text{DE-AEFA}}$); then initialize other parameters in DE, AEFA.

Step 2: Randomly initialize the values for the different parameters in a population of DE, AEFA individually.

Step 3: Evaluate fitness of respective populations and choose global solution whose population fitness value is highest.

Step 4: From Step-3, population having highest fitness value is survived and wipe out other individuals.

Step 5: The searching of DE and AEFA is carried on to the survived individuals as procedure in Step-4.

- **AEFA phase:** Particles positions and velocities are updated as given in Anita and Yadav (2019).
- **DE phase:** Mutation, Recombination and selection operations are performed.

Step 6: Scrutinize the derived solutions. If convergence criteria is met, terminate iterations and display best solution, or else go to Step-3.

$$f_{(x_1,x_2)} = \left(x_1^2 + x_2 - 11\right)^2 + \left(x_1 + x_2^2 - 7\right)^2 \qquad (1.22)$$

$$f_{(x_1,x_2)} = \left(x_1 + 2x_2 - 7\right)^2 + \left(2x_1 + x_2 - 5\right)^2 \qquad (1.23)$$

Moreover, the efficacy of presented algorithm is deliberated by testing on standard benchmark test functions like Himmelblau given in Equation 1.22 and booth function given in Equation 1.23, and the corresponding convergence characteristics are compared in Figures 1.12 and 1.13, respectively. Observing the convergence characteristics, it is found that the function value for the presented DE-AEFA starts at a good value and the best final solution is obtained in less number of iterations (Figure 1.12 and Figure 1.13). The optimal function values of the Himmelblau and Booth functions are noted in Tables 1.1 and 1.2, respectively. The procedural flow of presented DE-AEFA algorithm is shown in Figure 1.14.

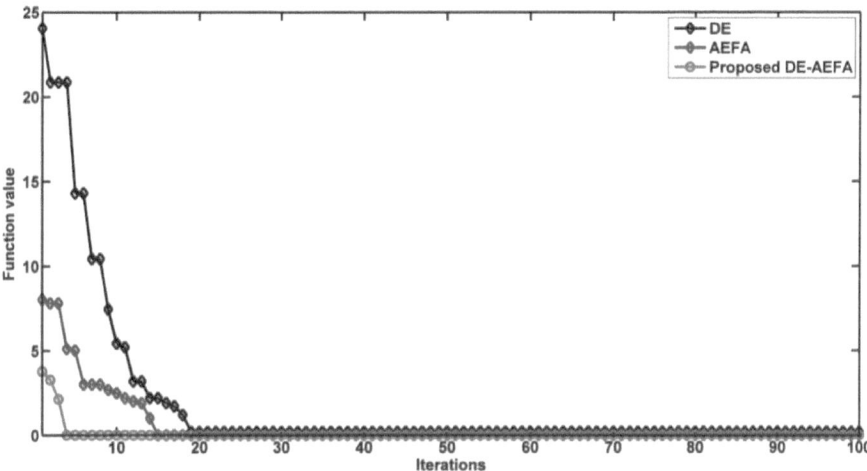

FIGURE 1.12 Convergence characteristics for Himmelblau function.

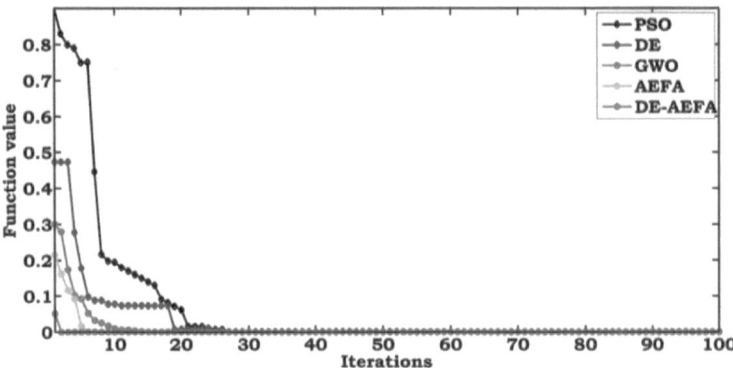

FIGURE 1.13 Convergence characteristics for Booth function.

TABLE 1.1
Optimal Solutions Obtained for Himmelblau Function

S. No	Parameters	DE	AEFA	DE-AEFA
1	X1	3.002461	2.993744	2.999176
2	X2	1.982402	2.004345	2.000937
3	Function value	0.00458	0.001223	2.46E-05

TABLE 1.2
Optimal Solutions Obtained for Booth Function

S. No	Parameters	PSO	DE	GWO	AEFA	DE-AEFA
1	X1	0.997417	0.998793	1.001176	1.000733	0.999995
2	X2	3.002246	3.000602	2.998958	2.999194	3.000000
3	Function value	1.22E-05	3.28E-06	2.54E-06	1.21E-06	1.43E-10

1.7 SIMULATION RESULTS

Analyses of both the test systems are carried out under the supervision of classical PID controller optimized with various algorithms for area-1 subjugated with 1%SLP. To get the investigation in a comparative manner the number of populations and maximum number of iterations are considered as 100 for every optimization algorithm. Optimization algorithms are developed in (.mfile) format and the test systems are developed in MATLAB/SIMULINK domain.

1.7.1 TEST SYTEM-1 DYNAMIC ANALYSIS

To showcase the effective functioning of the presented DE-AEFA based control technique, its performance is tested on the conventional power system model of two-area thermal unit with reheat turbine structure. The responses of the system under

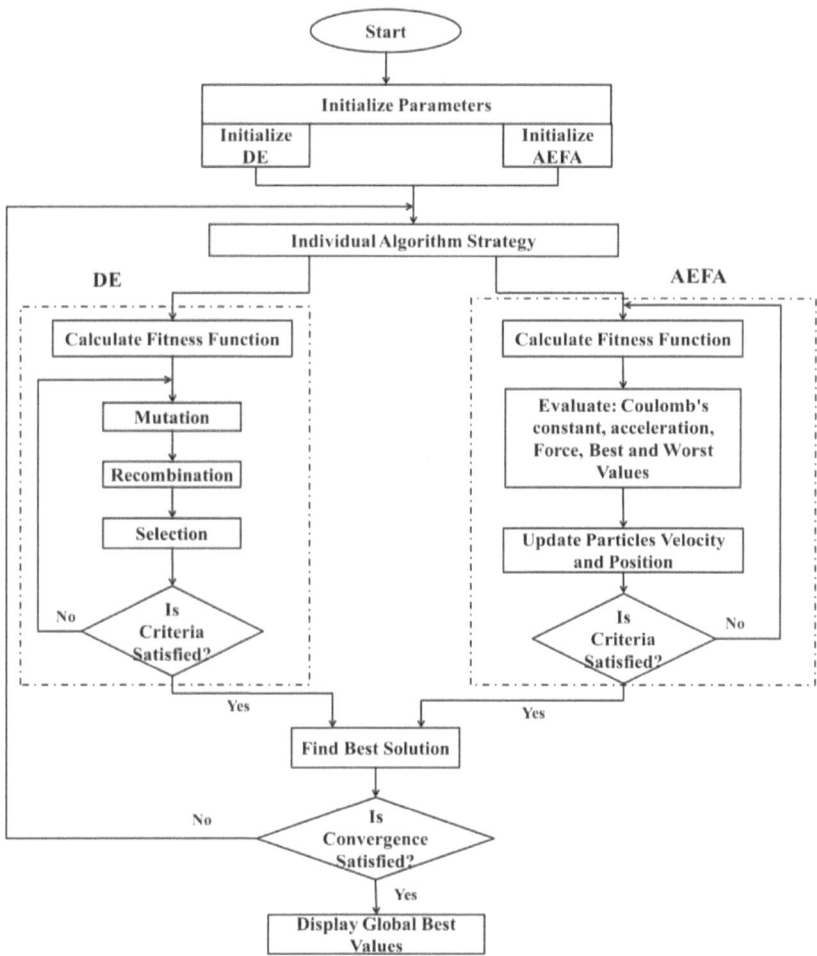

FIGURE 1.14 Flowchart of DE-AEFA algorithm.

DE-AEFA regulation is compared with other optimizations like PSO (Magid and Abido 2003), ABC (Gozde et al. 2012) and BFOA (Ali and Elazim 2013) algorithms reported in the literature and is shown in Figure 1.15, and the responses are numerically analyzed in point of view of settling time (T_s) given in Table 1.3. Examining Figure 1.15 and Table 1.3, it is revealed that the system behavior is strictly regulated under load disturbances with proposed control technique.

1.7.2 Test System-2 Dynamic Analysis

1.7.2.1 Performance Analysis of Optimization Algorithms

The effect of the PID controller on a system with LFC and AVR loops in each area is analyzed. The parameters of the controller are optimized with DE, AEFA and DE-AEFA algorithms one at a time. In area-1 for each instance, 1% SLP is impressed to

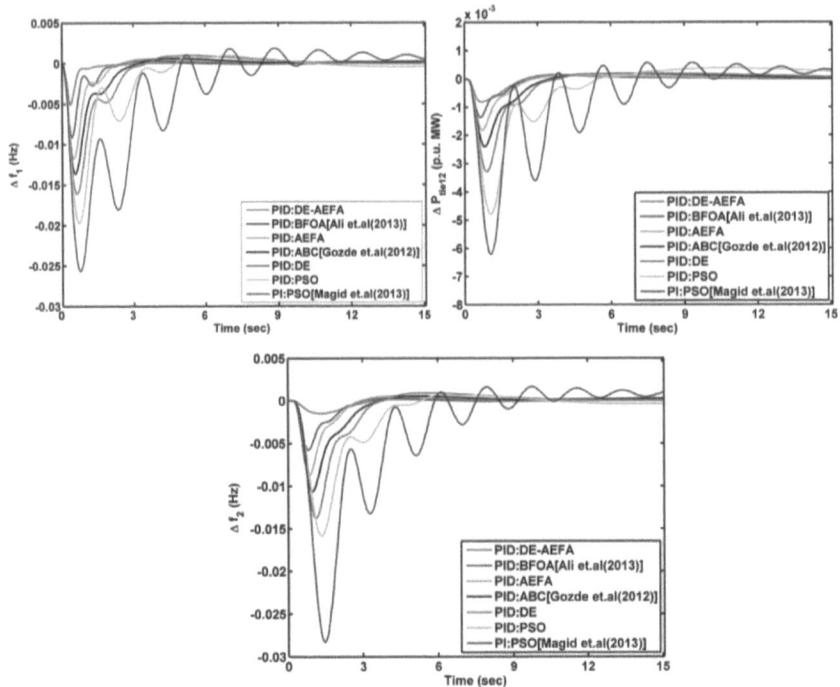

FIGURE 1.15   Test system-1 responses for area-1 with 1%SLP.

TABLE 1.3

Settling Time of Test System-1 Responses and Optimal Controller Gains

				Controller Optimization			
		PID: BFOA		**PID: ABC**			**PI: PSO**
	PID:	**(Ali and**	**PID:**	**(Gozde et**	**PID:**	**PID:**	**(Magid and**
Parameters	**DE-AEFA**	**Elazim 2013)**	**AEFA**	**al. 2012)**	**DE**	**PSO**	**Abido 2003)**
K_{P1}	1.967	1.746	1.544	1.337	0.926	0.919	0.757
K_{P2}	1.892	1.767	1.641	1.246	0.938	1.002	0.879
K_{I1}	0.019	0.044	0.035	0.005	0.055	0.020	0.306
K_{I2}	0.013	0.041	0.009	0.019	0.001	0.019	0.301
K_{D1}	1.062	0.721	0.907	0.914	0.729	0.586	-
K_{D2}	1.142	0.763	1.091	0.853	0.692	0.574	-
ISE*10^{-3}	0.498	1.143	2.475	5.433	11.046	27.69	63.78
Settling time: Δf_1	4.96	6.38	8.02	8.75	12.67	20.05	25.49
Settling time: ΔP_{tie}	6.14	7.83	10.57	12.05	17.30	23.89	26.27
Settling time: Δf_2	4.67	6.35	9.55	11.20	13.03	20.45	23.03

study the system dynamic response. The variations of dynamic responses with different optimization algorithms are depicted in Figure 1.16 and these are analyzed in terms of settling time (T_S), maximum undershoot (MUS) and maximum overshoot (MOS). The numerical results are noted in Table 1.4. From graphical and numerical results it is observed that the functioning of the PID controller with combined DE-AEFA algorithm is more superior to other optimization methods. The convergence characteristics of all optimization algorithms employed in this work are compared in Figure 1.17. Thus, further analysis is carried out through the DE-AEFA-based PID controller.

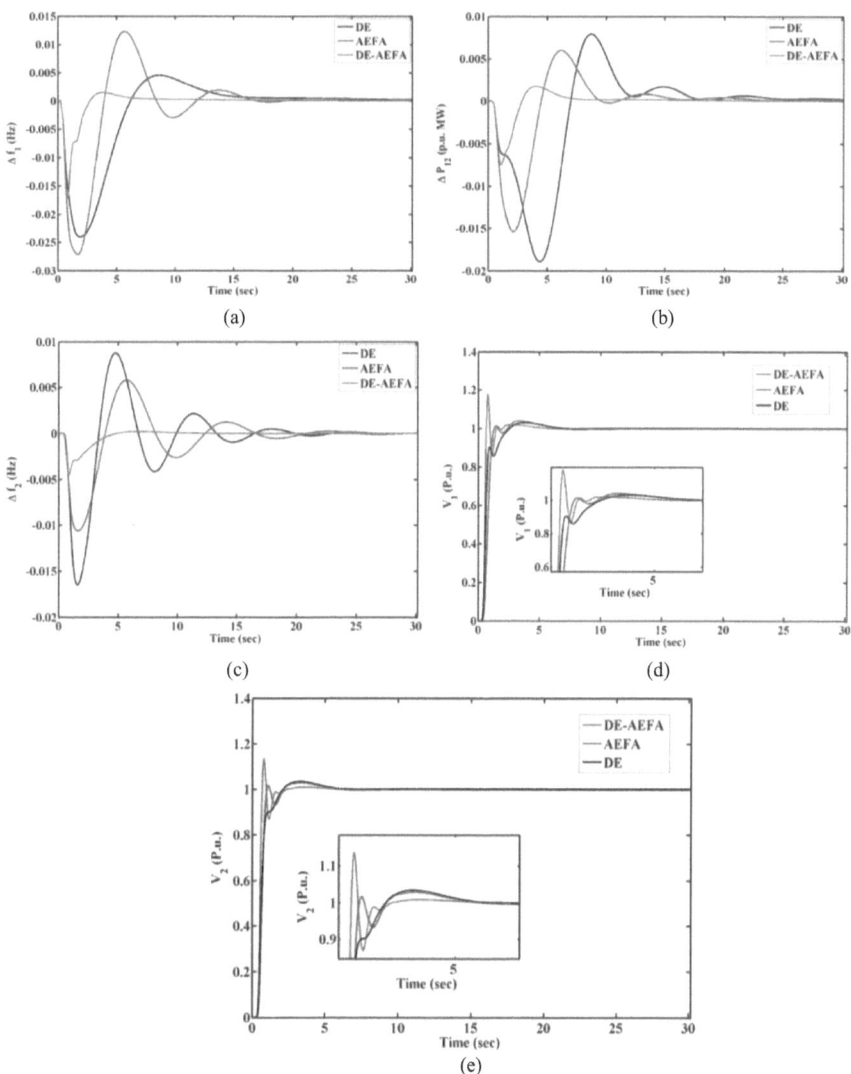

FIGURE 1.16 Responses of combined system (test system-2) under optimization algorithms based PID controller. (a) Δf_1, (b) $\Delta P_{tie1,2}$, (c) Δf_2, (d) V_1, (e) V_2.

TABLE 1.4

Numerical Results of Combined Model (Test System-2) for Different Optimization Algorithms

Parameters and Controllers		DE: PID	AEFA: PID	DE-AEFA: PID
T_S (s)	Δf_1	25.770	22.540	14.340
	Δf_2	25.75	23.26	10.52
	ΔP_{tie12}	28.85	23.20	18.43
	V_1	7.011	6.615	5.058
	V_2	8.982	7.920	4.629
MUS (P.u.)	$\Delta f_1 \times 10^{-3}$	−24.050	−27.120	−16.190
	$\Delta f_2 \times 10^{-3}$	−16.490	−10.620	−4.397
	$\Delta P_{tie12} \times 10^{-3}$	−18.920	−15.360	−7.426
	V_1	0.860	0.976	0.8805
	V_2	0.925	0.944	0.8693
MOS (P.u.)	$\Delta f_1 \times 10^{-3}$	4.567	12.320	1.462
	$\Delta f_2 \times 10^{-3}$	8.796	5.414	0.213
	$\Delta P_{tie12} \times 10^{-3}$	7.947	6.013	1.774
	V_1	1.030	1.037	1.181
	V_2	1.031	1.034	1.136
ISE		25.925	23.487	12.641

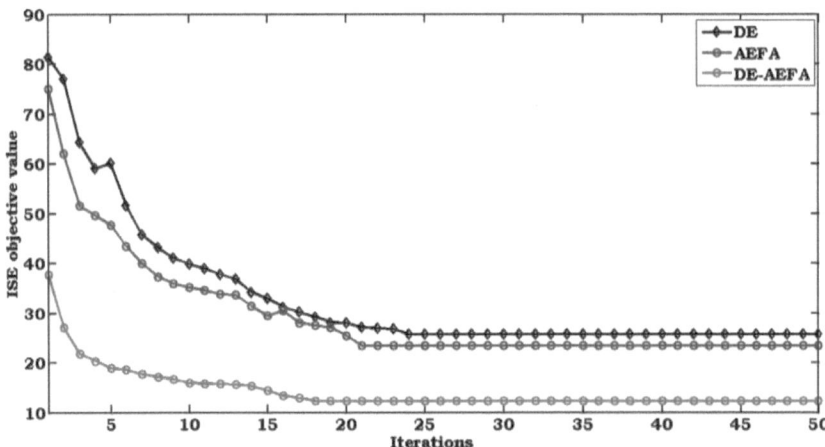

FIGURE 1.17   Convergence characteristics of optimization algorithms.

1.7.2.2   Effect of AVR Coupling on LFC

To manifest the impact of AVR coupling, the considered interconnected system is simulated with and without AVR coupling consideration provided with DE-AEFA based PID as secondary controller using ISE as performance index. The dynamic responses are compared in Figure 1.18 for 1% SLP enforced in area-1 for both the

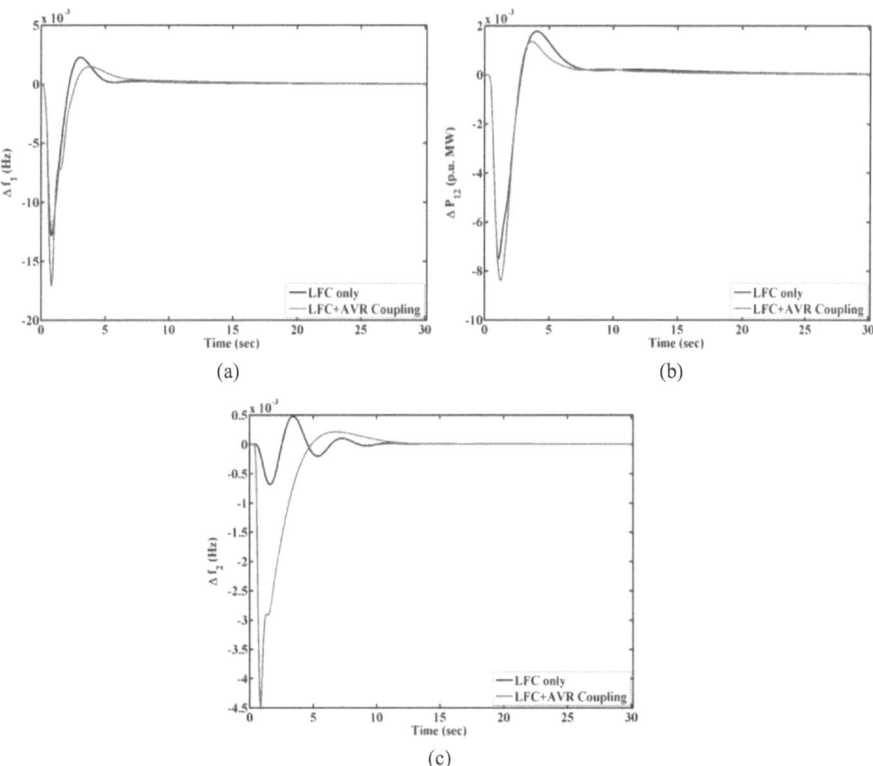

FIGURE 1.18 Comparison of test system-2 responses without and with AVR coupling under DE-AEFA based PID controller. (a) Δf_1, (b) $\Delta P_{tie1,2}$, (c) Δf_2.

cases. From Figure 1.18 it is clear that system frequency and tie-line power switching are greatly influenced by coupling AVR loop. From Equation 1.1 it is absolved that regulating terminal voltage through change in generator field excitation can maintain the frequency deviation. Thus, these two parameters which are to be regulated in order to maintain stability must be addressed aggregately.

1.7.2.3   Comparative Performance Evaluation of Various FACTS Devices

In this subsection, the test system under investigation is rendered with FACTS devices connected in series with the tie-line one at a time. The coordinated performance of these devices with proposed control methodology is put under comparison for 1% SLP impressed on area-1. Dynamic responses of the system while considering various FACTS devices are compared in Figure 1.19 and the numerical data is mentioned in Table 1.5. From Figure 1.19 and Table 1.5 it is observed that system frequency deviations in both the areas, tie-line power variations are magnificently controlled and the generator terminal voltages are maintained at prescribed value in short duration by the proposed controller with IPFC device compared to SSSC, TCPS, and TCSC devices. Thus, IPFC device is declared as the best damping or frequency controller among SSSC, TCPS and TCSC for combined LFC and AVR

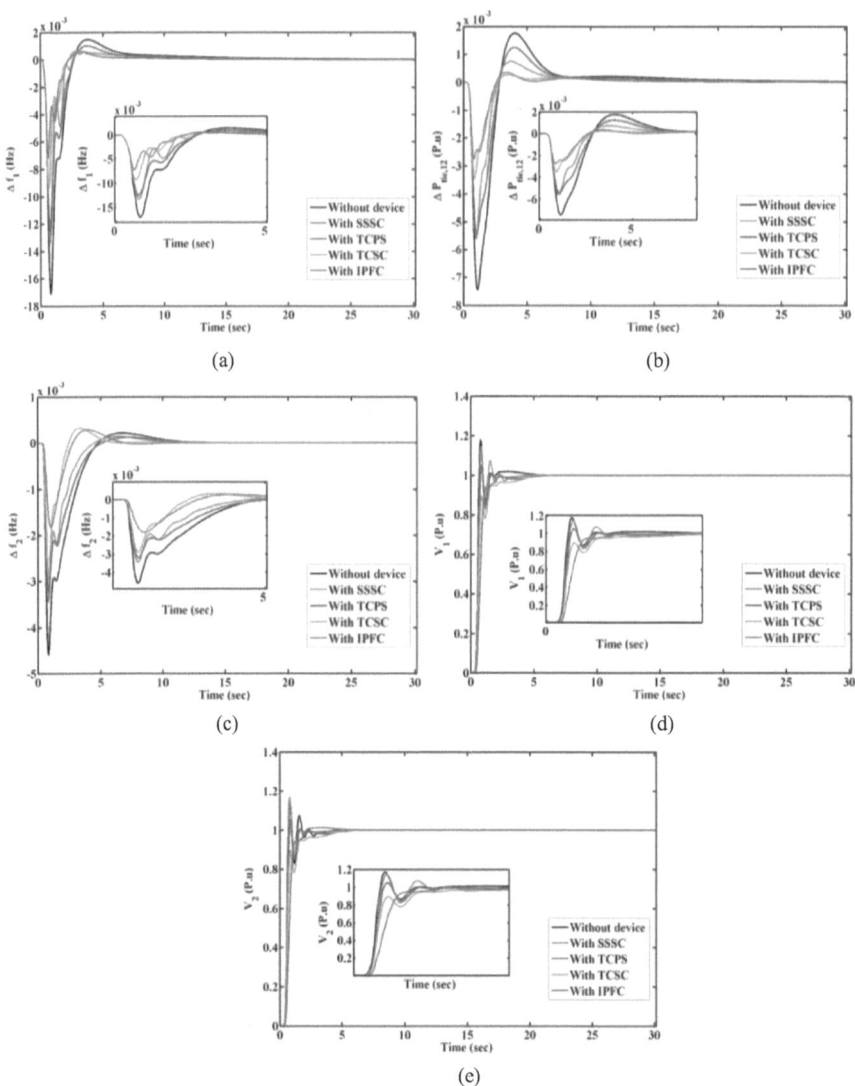

FIGURE 1.19 Performance comparison of various FACTS devices in the combined model. (a) Δf_1, (b) $\Delta P_{tie1,2}$, (c) Δf_2, (d) V_1, (e) V_2.

loops. The controller parameters and parameters of FACTS devices are placed in Table 1.6. For easy analysis the settling times of dynamic responses under various FACTS devices are depicted in Figure 1.20.

1.7.2.4 Coordinated Control Strategy of IPFC-RFBs

Furthermore, to make the control strategy for the combined model system for simultaneous stabilization of frequency and terminal voltage more sovereign, the coordinated control strategy of IPFC-RFBs is implemented. Figure 1.21 depicts the

TABLE 1.5

Numerical Results of Combined Model (Test System-2) for Incorporation of Various FACTS Devices

Parameters and Proposed Controller with FACTs Devices		Without Device	With SSSC	With TCPS	With TCSC	With IPFC
T_S (s)	Δf_1	14.340	12.704	11.347	10.812	8.481
	Δf_2	10.520	9.970	7.574	5.858	6.187
	ΔP_{tie12}	18.430	16.522	14.002	12.130	10.350
	V_1	5.058	4.997	4.343	4.247	3.519
	V_2	4.629	4.602	4.251	4.143	3.484
MUS (P.u.)	$\Delta f_1 \times 10^{-3}$	−16.190	−13.030	−12.560	−9.138	7.071
	$\Delta f_2 \times 10^{-3}$	−4.397	−3.414	−3.196	−2.900	−1.805
	$\Delta P_{tie12} \times 10^{-3}$	−7.426	−5.592	5.405	−3.380	−2.717
	V_1	0.8805	0.836	0.858	0.7920	0.9387
	V_2	0.8693	0.849	0.858	0.7913	0.9299
MOS (P.u.)	$\Delta f_1 \times 10^{-3}$	1.462	0.588	1.029	0.555	0.538
	$\Delta f_2 \times 10^{-3}$	0.213	0.141	0.375	0.315	0.289
	$\Delta P_{tie12} \times 10^{-3}$	1.774	0.744	1.242	0.269	0.3517
	V_1	1.181	1.131	1.051	1.000	1.000
	V_2	1.136	1.130	1.049	1.000	1.000
ISE		12.641	9.814	7.905	5.126	3.913

TABLE 1.6

PID Controller Optimal Gains and Parameters of FACTS Devices Considered for Combined Model (Test System-2)

	Area-1		Area-2		
Controller	LFC Loop	AVR Loop	LFC Loop	AVR Loop	FACTS Device Constants
DE-AEFA based PID without FACTS	$K_P=3.8147$ $K_I=2.9058$ $K_D=1.1270$	$K_P=2.0975$ $K_I=1.6324$ $K_D=0.9134$	$K_P=3.2785$ $K_I=2.5469$ $K_D=1.9575$	$K_P=1.9649$ $K_I=1.5156$ $K_D=0.9706$	-
DE-AEFA based PID with SSSC	$K_P=3.9572$ $K_I=2.4858$ $K_D=1.8003$	$K_P=2.1494$ $K_I=1.4298$ $K_D=0.9157$	$K_P=3.7922$ $K_I=2.9595$ $K_D=1.6557$	$K_P=2.0357$ $K_I=1.8491$ $K_D=0.9340$	$T_1=0.0046$, $T_2=0.0540$ $T_3=0.0490$, $T_4=0.0630$ $K_{SSSC}=1$, $T_{SSSC}=0.9102$
DE-AEFA based PID with TCPS	$K_P=4.0786$ $K_I=2.7577$ $K_D=2.0431$	$K_P=2.3922$ $K_I=1.6555$ $K_D=1.1712$	$K_P=3.7060$ $K_I=3.0318$ $K_D=2.2769$	$K_P=2.0462$ $K_I=2.0971$ $K_D=1.8235$	$K_\varphi=1.1835$ $T_{PS}=0.9390$
DE-AEFA based PID with TCSC	$K_P=3.9638$ $K_I=2.3171$ $K_D=1.9502$	$K_P=1.7655$ $K_I=2.1869$ $K_D=0.7952$	$K_P=4.0344$ $K_I=3.4387$ $K_D=2.3816$	$K_P=2.4898$ $K_I=2.4456$ $K_D=1.6463$	$T_1=0.0605$, $T_2=0.0503$ $T_3=0.0483$, $T_4=0.1068$ $K_{TCSC}=0.9970$, $T_{TCSC}=0.973$

(Continued)

TABLE 1.6 (*Continued*)

PID Controller Optimal Gains and Parameters of FACTS Devices Considered for Combined Model (Test System-2)

	Area-1		Area-2		
Controller	LFC Loop	AVR Loop	LFC Loop	AVR Loop	FACTS Device Constants
DE-AEFA based PID with IPFC	$K_P=4.7094$ $K_I=2.7547$ $K_D=2.2760$	$K_P=2.6797$ $K_I=1.6551$ $K_D=1.6126$	$K_P=3.4984$ $K_I=2.1190$ $K_D=1.9597$	$K_P=2.3404$ $K_I=1.5853$ $K_D=1.2238$	$K_1=0.9174, K_2=0.9452$ $T_{IPFC}=0.8256$
DE-AEFA: PID with RFBs only	$K_P=4.6983$ $K_I=2.1317$ $K_D=1.9620$	$K_P=1.7751$ $K_I=2.1906$ $K_D=0.8502$	$K_P=4.0214$ $K_I=3.3378$ $K_D=2.3791$	$K_P=2.4988$ $K_I=2.4557$ $K_D=1.6560$	$K_{RBF1}=1.017, K_{RBF2}=0.903$ $T_{RBF1}=0.999, T_{RBF2}=1.000$
DE-AEFA: PID with IPFC-RFBs	$K_P=4.8141$ $K_I=2.6573$ $K_D=2.2913$	$K_P=2.7727$ $K_I=1.6615$ $K_D=1.5962$	$K_P=3.9448$ $K_I=2.5180$ $K_D=1.9787$	$K_P=2.4043$ $K_I=1.8551$ $K_D=1.3273$	$K_1=0.971, K_2=0.957$ $T_{IPFC}=0.9265,$ $K_{RBF1}=0.9070,$ $K_{RBF2}=0.9351,$ $T_{RBF1}=0.997, T_{RBF2}=0.9876$

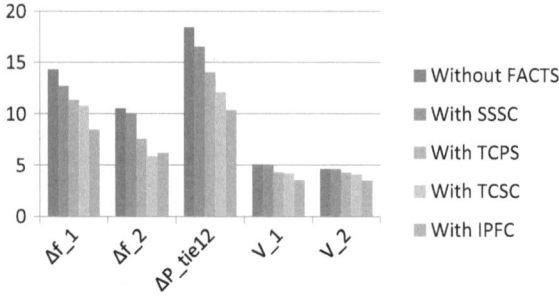

FIGURE 1.20 Comparison of settling time (in s) for dynamic responses under considering various FACTS devices.

TABLE 1.7

Numerical Results of Combined Model (Test System-2) with IPFC-RFBs Devices

	Parameters	Without Device	With RFBs Only	With IPFC Only	With IPFC-RFBs
T_S (s)	Δf_1	14.340	12.27	8.481	6.102
	Δf_2	10.520	8.19	6.187	6.067
	ΔP_{tie12}	18.430	15.71	10.350	9.541
	V_1	5.058	4.912	3.519	3.497
	V_2	4.629	4.012	3.484	3.382

(*Continued*)

TABLE 1.7 (*Continued*)
Numerical Results of Combined Model (Test System-2) with
IPFC-RFBs Devices

Parameters		Without Device	With RFBs Only	With IPFC Only	With IPFC-RFBs
MUS (P.u.)	$\Delta f_1 \times 10^{-3}$	−16.190	−9.734	7.071	−5.853
	$\Delta f_2 \times 10^{-3}$	−4.397	−3.016	−1.805	−1.779
	$\Delta P_{\text{tie}12} \times 10^{-3}$	−7.426	−3.520	−2.717	−2.200
	V_1	0.8805	0.9032	0.9387	0.945
	V_2	0.8693	0.86893	0.9299	0.9484
MOS (P.u.)	$\Delta f_1 \times 10^{-3}$	1.462	0.8349	0.538	0.496
	$\Delta f_2 \times 10^{-3}$	0.213	0.154	0.289	0.263
	$\Delta P_{\text{tie}12} \times 10^{-3}$	1.774	0.635	0.3517	0.205
	V_1	1.181	1.130	1.000	1.000
	V_2	1.136	1.135	1.000	1.000
ISE		12.641	9.070	3.913	0.097

responses of the combined model system for the case of without considering any devices, with IPFC consideration in tie-line, with placing RFBs in area 1 and 2 and with considering both IPFC and RFBs. The numerical data related to the system responses of this case are noted in Table 1.7. Observing Figure 1.21 concluded that the system responses under load disturbances are greatly enhanced with the coordinated control of IPFC-RFBs under DE-AEFA-based PID controller.

1.8 CONCLUSION

In this chapter, efforts have been made to assess the performance of various FACTS controllers in combined LFC and AVR of multi-area interconnected power systems. Initially the performance of presented DE-AEFA searching algorithm is tested by applying on the conventional power system model of dual area thermal systems having reheat turbine structure. Investigation showcased the superiority of DE-AEFA approach over other control approaches reported in the literature. Later, it is extended to the combined model having PID as controller for both LFC and AVR loops. Dynamic responses for 1%SLP in area-1 reveals the effectiveness of DE-AEFA based controller over others. Later, the impact of AVR coupling on LFC loop is demonstrated and is justified. Finally, the system under investigation is provided with various FACTS devices in series with the tie-line one at a time and the dynamic responses are studied. Comparison of the coordinated performances of FACTS devices with proposed controller brings out the predominant functioning of IPFC as damping controller compared to TCSC, TCPS and SSSC. Finally, the coordinated control strategy of IPFC-RFBs is implemented in the combined model and drives the conclusion that the coordinated performance of IPFC-RFBs are highly dominating in damping out the system deviations further.

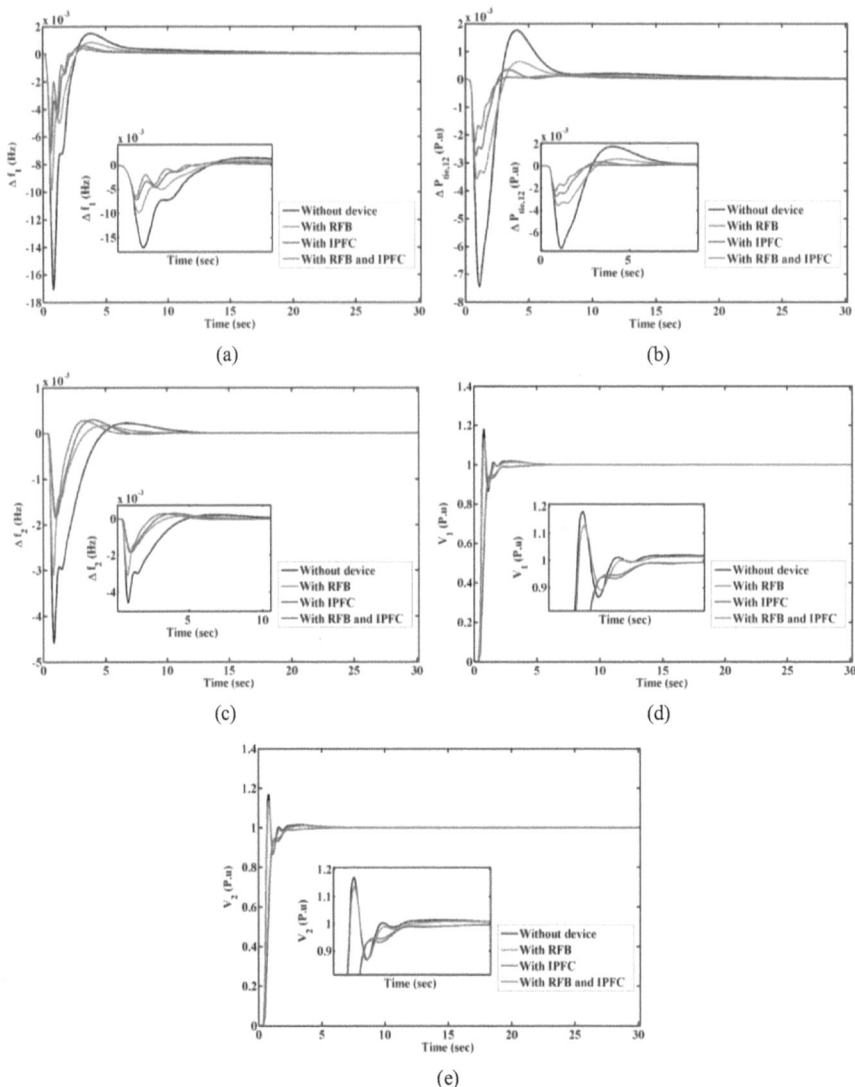

FIGURE 1.21 Responses of combined model (test system-2) with the coordinated IPFC-RFB control strategy. (a) Δf_1, (b) $\Delta P_{tie1,2}$, (c) Δf_2, (d) V_1, (e) V_2.

REFERENCES

Ali, E.S., and S.M. Abd-Elazim. 2013. "BFOA based design of PID controller for two area load frequency control with nonlinearities". *Electrical Power and Energy Systems* 51: 224–231.

Anita, and A. Yadav. 2019. "Artificial electric field algorithm for global optimization". *Swarm and Evolutionary Computation Base Data*, https //doi.org/10.1016/j.swevo.2019.03.013.

Arya, Y. 2019. "A new optimized fuzzy FOPI-FOPD controller for automatic generation control of electric power systems". *Journal of Franklin Institute* 356: 5611–5629.

Bheem, S., and K. Deepak. 2019. "Dual loop IMC structure for load frequency control issue of multi-area multi-sources power systems". *Electrical Power and Energy Systems* 112: 476–494.

Chandrakala, K.R.M.V., and S. Balamurugan. 2016. "Simulated annealing based optimal frequency and terminal voltage control of multi source multi area system". *Electrical Power and Energy Systems* 78: 823–829.

Fathy, A., and A.M. Kaseem. 2018. "Antlion optimizer-ANFIS load frequency control for multi-interconnected plants comprising photovoltaic and wind turbine". *ISA Transactions*, https://doi.org/10.1016/j.isatra.2018.11.035.

Fosha, E., and O. Elgerd. 1970. "The megawatt-frequency control problem: A new approach via optimal control theory". *IEEE Transactions on Power Apparatus and Systems PAS* 89 (4): 563–577.

Gozde, H., M.C. Taplamacioglu and I. Kocaarslan. 2012. "Comparative performance analysis of artificial bee colony optimization in automatic generation control for interconnected reheat thermal power system". *Electrical Power and Energy Systems* 42: 167–178.

Guha, D., P.K. Roy, and S. Banerjee. 2016. "Load frequency control of large scale power system using quasi-oppositional grey wolf optimization algorithm". *Engineering Science and Technology, an International Journal* 16: 1693–1713.

Lal, D.K., and A.K. Barisal. 2017. "Comparative performances evaluation of FACTS devices on AGC with diverse sources of energy generation and SMES". *Cogent Engineering* 4: 1318466.

Magid, Y.L.A., and M.A. Abido. 2003. "AGC tuning of interconnected reheat thermal systems with particle swarm optimization". *IEEE International Conference on Electronics, Circuits and Systems* 01: 376–379.

Rajbongshi, R., and L.C. Saikia. 2017. "Combined control of voltage and frequency of multi-area multisource system incorporating solar thermal power plant using LSA optimized classical controllers". *IET Generation, Transmission & Distribution* 11 (10): 2489–2498.

Rajbongshi, R., and L.C. Saikia. 2018. "Combined voltage and frequency control of a multi-area multisource system incorporating dish-Stirling solar thermal and HVDC link". *IET Renewable Power Generation* 12 (03): 323–334.

Rakshani, E., K. Rouzbhei, and Sadeh. 2009. "A new combined model for simulation of mutual effects between LFC and AVR loops". *Asia-Pacific Power and Energy Engineering Conference*, Wuhan 1–5.

Shankar, R., C. Kalyan, and B. Ravi. 2016. "Impact of energy storage system on load frequency control for diverse sources of interconnected power system in deregulated power environment". *Electrical Power and Energy Systems* 79: 11–26.

Storn, R., and K. Price. 1997. "Differential evolution – A simple and efficient heuristic for global optimization over continuous spaces". *Journal of Global Optimization* 11 (04): 341–359.

Tasnin, W., and L.C. Saikia. 2018. "Comparative performance of different energy storage devices in AGC of multi-source system including geothermal power plant". *Journal of Renewable and Sustainable Energy*, https://doi.org/10.1063/1.5016596.

2 Reliable Energy Storage System for the Grid Integration

Kummara Venkata Guru Raghavendra, and Kisoo Yoo
Yeungnam University

CONTENTS

2.1 INTRODUCTION

Rapid population growth is leading to a huge demand for electricity generation. Energy production and consumption are mainly based on fossil fuels, which have a dominant impact on the world economy and nature. Therefore, the demand for energy from renewable sources is the last. Energy storage is an important component of renewable energies. This is why high-quality renewable energy storage elements

DOI: 10.1201/9781003143802-2

are of particular interest to researchers and manufacturers. According to the principles of electrical and chemical energy conversion, the most important energy storage devices are SCs, fuel cells, and batteries. Recently, the SC has received much attention due to its unique properties, such as higher C, longer service life, and improvement (Ps). Besides, SCs are easy to use, require less maintenance, do not suffer from memory effects, and are safe with greater power and strength between a large capacitor (PS) and potential batteries and device [1–4] is allowed. This has proven to be a great solution for power supply and remote zones where access to electricity is almost risky and expensive. SCs are currently used in many compact power supplies, such as laptops, cell phones, cameras, and more. They are compact and flexible.

In some cases, SCs have different applications for electric and hybrid vehicles to deal with the energy needed for acceleration during braking, to save energy, and to protect batteries (with high frequency in billing processes and quickly empty, i.e. during unstable operation). A wagon diagram is shown in Figure 2.1 displaying the special power of Ps and the special energy of Ito. The graph describes the improvement of fuel cells; instead, SCs have higher PS. Even batteries are average in terms of energy and energy potential. The framework also illustrates that the PS and E of SCs overlap in fuel cells and batteries. Figure 2.1 shows that there is no electrochemical device comparable to an internal combustion engine. It is, therefore, necessary to improve the electrochemical systems Ps and E to drive the size of the internal combustion engines [5].

Leyden Jar in the 18th century, has depicted the origin of SCs, earlier they were made of metal foil, here metal film was used as either electrode or dielectric. At the time of charging, both charges (positive and negative) were accumulated at the corresponding electrodes.

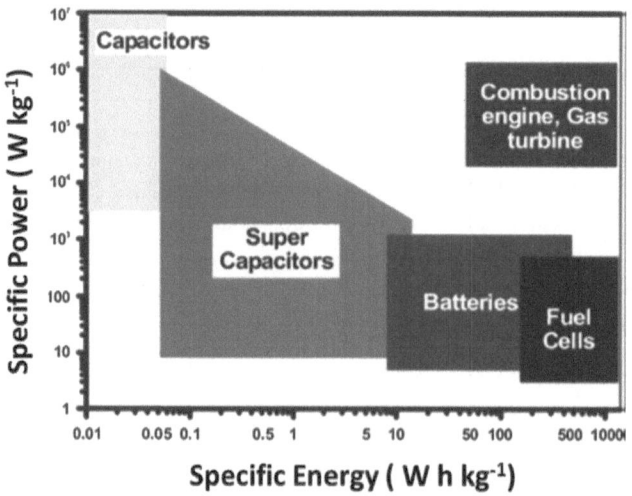

FIGURE 2.1 Ragone plots of different electrochemical energy conversion systems (EECS), the combustion engines, the turbines, and the traditional conventional capacitors [5]. (Adapted with permission from Ref. [5] Copyright American Chemical Society (2004).)

In the early 1920s, the first electrolytic capacitor took its shape. General Electric developed its first SC of EDLC type that acted as capacitor plates in 1957. The phenomenon of electrostatic storage occurs in EDLCs, i. A charge accumulates between the electrode and the electrolyte, which makes the stability of its more flexible. EDLC structures vary, as do carbon materials such as carbon nanotubes (CNTs), activated carbon (AC), carbon derived compounds (CDCs), graphene, and more. The unique properties of these EDLCs are: they have a better specific surface area (SSA), better mechanical and chemical stability; they also have good electrical conduction. To increase SC Cs, new electrochemically active materials for pseudo-capacitors have been investigated for use in delayed transition phenomena.

RuO$_2$ is widely studied by B.E. Conway from 1975 to 1980 as a pseudocondensing material. The phenomenon of absorbing the load of this type of SC is based on electrosorption, intercalation, oxidation-red reaction [6]. This phenomenon of disappearing processes allows pseudo-capacitors to have higher capacity and higher E compared to EDLCs. The strength of pseudo-capacitors is derived from the phenomenon of electron charge transfer on the electron-electrolyte interface, obtained from dissolved and adsorbed ions. There is no reaction to these announced ions following the electrode material because only charge transfer happens.

The inclination of an electrode to achieve pseudo-capacitance properties depends on the chemical correlation of the electrode material for the absorption of ions on the surface of the electrode with the structure, as well as the size of the electrode pore. As the applied voltage increases, there is a linear improvement in storage. Transition metal oxides (TMO), such as IrO$_2$, RuO$_2$, V$_2$O$_5$, Fe$_3$O$_4$, Co$_3$O$_4$, etc. are interesting materials today in pseudocapacitor applications. Conductive polymers (CP) and transfers of metal sulfides such as polyaniline (PANI), polythiophene, poly (3,4-ethylene dioxythiophene) (PEDOT), polyvinyl alcohol (PVA), poly (4-styrene sulfonate) (PSS), polyacetylene, etc. are also high-performance materials.

In 2007, FDK developed the first lithium-ion capacitors, also called hybrid capacitors. These capacitors have a combination of carbon electrodes with a lithium-ion electrode, which leads to better Cs and reduces the anodic potential, which eventually increases the cell voltage and thus Es. In these types of setups, the faradate electrode has higher Cs, and on the contrary, an unbroken electrode offers a higher density. Therefore, recent studies have placed more emphasis on hybrid capacitor mixtures (obtained by carbon material in TMOs or CPs) and battery type (connected SC electrode with battery electrode). Table 2.1 shows a comparison of the analysis of the three types of SC based on different electrochemical parameters.

Traditional capacitors have some limitations, such as stiffness and mass nature. Therefore, they are not suitable for future applications. The fresh technology of SC, which allows new unique features, is more suitable for many electronic applications. Commonly used electroactive materials are activated carbon (AC) and transition metal oxides (TMO). The properties of these materials for electrodes are irregular morphological uniformity; very weak graphite frames obtained with wider pores, having a minimum conductivity of TMOs, which makes it difficult to keep up with energy velocities at high speeds [8]. To overcome these limitations, new materials are required for SC electrodes, such as covalent organic frames (COFs), organometallic frames (MOF), metal nitrides, MXenes, metal sulfides, 2-D materials, mixing of leads, and so on [9–11].

TABLE 2.1

Comparison EDLC, Pseudocapacitors and Hybrid SC [7]

Parameters	Electrochemical Double-Layer Capacitor (EDLC)	Pseudocapacitors	Hybrid Capacitor
Electrode material	Carbon	Metal oxides (MOs) and conducting polymers (CPs)	Both carbon and MOs/CPs
Charge storage phenomenon	Non-Faradaic	Faradaic	Both Faradaic and non-Faradaic
Specific capacitance	Low	High	Intermediary
Specific energy	Low	High	High
Rate capability	Good	Low	Good
Cycling stability	Good	Low	Moderate

The biggest challenge facing SC is less Cs. Because it is directly related to C and the voltage (V) is square, either the potential or both can be improved to increase the Es, which can be achieved by higher electrode material Cs with wider voltage and optimized integrated systems for Window Electrolyte. Although the synthesis of each component (e.g., electrode materials and electrolytes, etc.) in the SC appears to be simple, there is a great need for coordination between pore size as well as structural behavior. of the ion-sized electrode material to promote the combined effect of electrolyte.

The choice of electrolytes and solutions plays an important role in creating some useful properties such as Ps, conductivity and temperature range. Some other electrolyte requirements for SC are windows with higher potential (PW), good stability, higher ion concentration, lower equivalent resistance (ESR), less volatility and values of viscosity, non-toxic nature, less ionic value, and finally lower costs [12].

2.2 THE WORLDWIDE SC MARKET

SCs are used in many applications because they are a safe, environmentally friendly, and economical energy source for the industry. The global SC market is also driven by many factors, such as the increase in applications for regenerative braking systems for elevators and hybrid vehicles (HEVs), which are increasing demand for SC applications in energy cleaning and the use of SC in locomotives. such as trains and planes. The market for electrolytes, which are subject to regulations and violations, is also evaluated by studying the selection of electrolytes by SC manufacturers [13]. According to the IDTechEx report, the global market share is expected to reach $ 8.3 billion by 2025. This market is expected to grow by a combined 30% by 2025 (Figure 2.2) [13].

Also, Figure 2.3a describes the breakdown of the SC consumption base by industry, separated by income, which is usually divided into three levels. Figure 2.3b provides regional consumer information. The Asia-Pacific (APAC) regions describe the largest contribution and will continue to do so, with the growth rate in 2016–2022 inevitable. Improving startups in this area are leading to market growth. North American (NA) and European industries followed the APAC growth. There are many energy

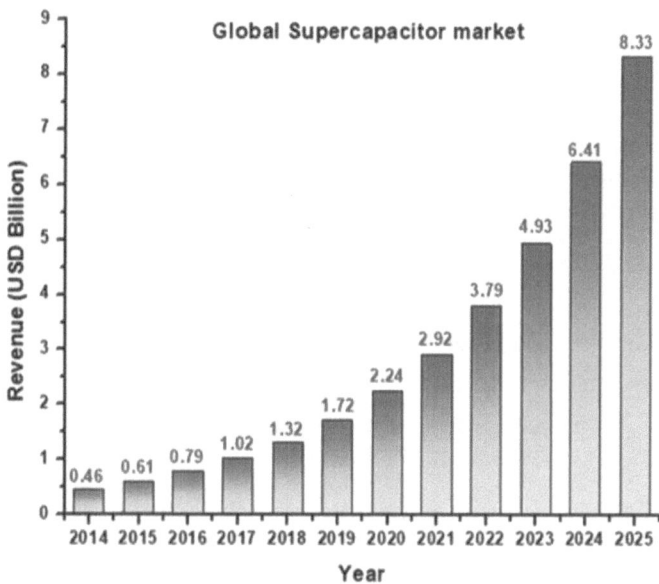

FIGURE 2.2 Global SC market.

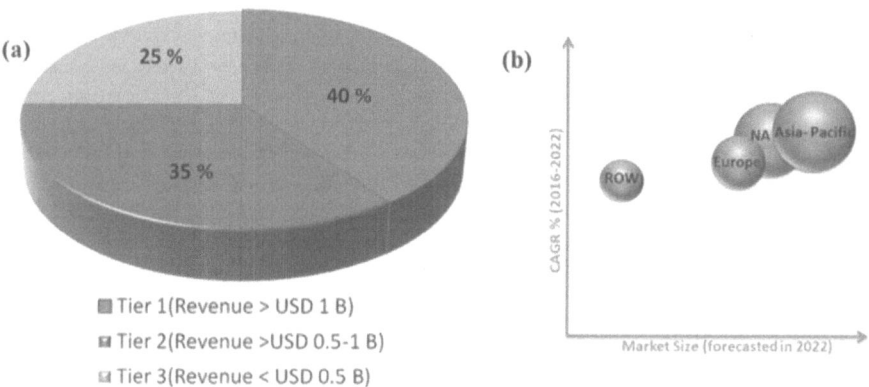

FIGURE 2.3 Consumption of SCs (a) based on the company type and (b) based on region.

storage systems and renewable energy projects in Japan, Singapore, South Korea, as well as in Australia. Government agencies in several Asian countries have also hybridized the transportation system and bring most APACs to the SC market. The price of the material is the main obstacle to market growth. Recently, some SCs have used graphene as a superconducting material, which is more expensive than activated carbon. Although such factors inhibit the growth of SCs in the market, large studies are being introduced to reduce overhead costs and help improve the acquisition rate of SCs in the current market [13].

2.2.1 Manufacturers

The number of SC developers and producers is growing rapidly. The main market is organic sodium chloride, which deals with propylene carbonate or acetonitrile electrolyte. However, electrolytes have recently provided almost 50% of the non-combustible and non-toxic electrolytic equipment available from the manufacturer in unique quality. Several SC applications vary depending on the needs of the ESS. The most important manufacturing facilities in the world are listed in Table 2.2 [14].

TABLE 2.2
Leading Manufacturing Companies of SCs [14]

Manufacturer	Country	Founded	Remarks
Cellergy	USA	2002	Manufactures the low cost electrochemical double layer capacitors (EDLC's), The main use of Cellergy SCs is in digital and portable electronic devices. The ultra-thin 1.7 mm packaging supports several small types of lightweight products.
Ioxus	USA	2007	Ioxus ultracapacitors are optimized for maximum performance and minimum resistance. They are operated in a wide range of temperatures, are lightweight, and designed for long-term use, and significantly increase the maximum power of the system in combination with large energy sources to optimize the product. Ninety five percent reusable, Ioxus ultracapacitors are the epitome of green technology.
Maxwell Technologies (now Tesla)	USA	1965	Maxwell's ultracapacitors are rapidly evolving and increasingly used technology that can charge and discharge energy quickly and efficiently. Thanks to many advantages, ultracapacitors are now used in different applications and are used for wide applications. Ultracapacitors supplement the main source of energy, that cannot usually cause rapid explosions, such as a combustion engine, battery, or fuel cell. The horizon of the future looks bright for ultracapacitors, that are readily distinguished as a potential alternative energy source.
Murata Manufacturing	Japan	1944	Murata offers various kinds of capacitors as silicon, ceramic, trimmer of polymer aluminum electrolyte, single-layer microchip.
Nanoramic Laboratories	USA	2008	Nanoramic Laboratories is the exclusive developer, manufacturer, and licensor of Neocarbonix electrodes dedicated to lithium-ion batteries, lithium-ion capacitors, and SCs. Their proprietary material and inexpensive processes to make Neocarbonix electrodes with an advanced three-dimensional nanoscopic carbon bond structure. The resulting product offers higher strength, specific energy, and performance under extreme conditions compared to traditional battery designs...

(Continued)

TABLE 2.2 (*Continued*)
Leading Manufacturing Companies of SCs [14]

Manufacturer	Country	Founded	Remarks
NEC Tokin (now Tokin-A KEMET group)	Japan	1938	A leading manufacturer of SCs and tantalum capacitors
Nippon Chemi-Con	Japan	1931	Nippon Chemi-Con offers electric double-layer capacitors, which are suitable for kinetic energy capture.
Panasonic	Japan	1918	Offers products such as film capacitors, aluminum electrolytic capacitors. Electric double layer capacitors (gold capacitor), conductive polymer electrolytic capacitors.
Paper Battery Company	USA	2008	The developer of novel SC-relied energy storage and battery technology,
Skeleton Technologies	Estonia	2009	Skeleton Technologies is the global leader in ultracapacitor & SC energy storage and helps companies in automotive, transportation, grid, and industrial applications.
Yunasko	UK	2010	Yunasko is a developer of highly efficient ultracapacitors with power and energy characteristics superior over existing competitors.
ZapGo	UK	2013	Leading producer of ultrafast charge and nanocarbon SCs for electrical vehicles, grid and many industrial applications.
SPSCap	China	2003	World-class manufacturer of ultra-capacitor products including cells, systems, and technological solutions.

2.3 CLASSIFICATION OF SC

SCs store energy based on two types of capacitive natures: bilayer electric force (EDL), obtained on the inner surface of the electrode with the rapid velocity of different electrostatic charges, and the electrolyte and pseudo-capacity formed by rapid redox and reversible processes. Different types of SC based on energy storage mechanisms, including electroactive material, are shown in Figure 2.4.

2.3.1 ELECTRICAL DOUBLE-LAYER (EDL) IN SC

The equation obtained by computational storage phenomena has a higher charge collection, a larger area, and acts as a limiting factor for conventional capacitors [15–18]. EDL-based SCs use a wider interface range as well as different isolation tasks for higher load potential. EDL is usually a structure in a particle that is loaded with liquid (Figure 2.5).

Theoretically, auxiliary silos are formed by balancing ions in a liquid and their concentration is close to the surface. Many different theories or models have been proposed in the literature study on this fluid and solid interface. The different models are shown in Figure 2.6 and each model is conveniently described.

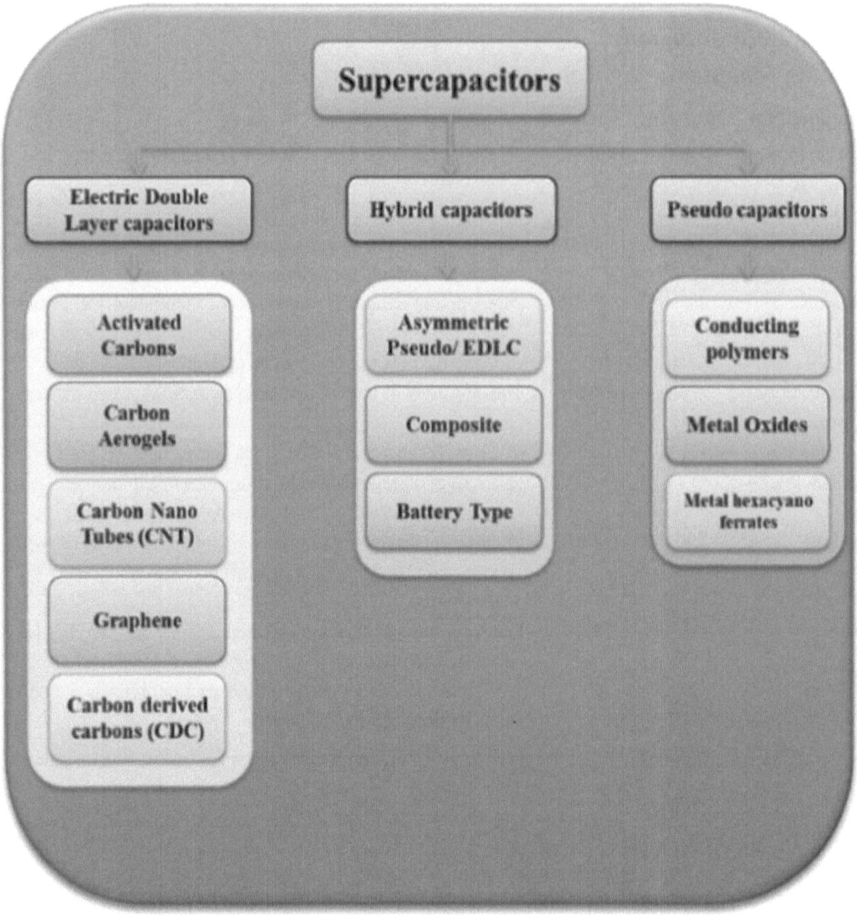

Supercapacitors
Electric Double Layer capacitors
Hybrid capacitors
Pseudo capacitors
Activated Carbons
Carbon Aerogels
Carbon Nano Tubes (CNT)
Graphene
Carbon derived carbons (CDC)
Asymmetric Pseudo/ EDLC
Composite
Battery Type
Conducting polymers
Metal Oxides
Metal hexacyano ferrates

FIGURE 2.4 Classification of SCs.

(a)
Electrolyte
Separator
Anode
Cathode
Anion
Cation
Solvent molecule
(b)
Electrolyte
Separator
Anode
Cathode
Metal oxide or redox active molecule
(c)
Electrolyte
Separator
Anode
Li⁺
Cathode
Li-based salt
Li⁺ ions

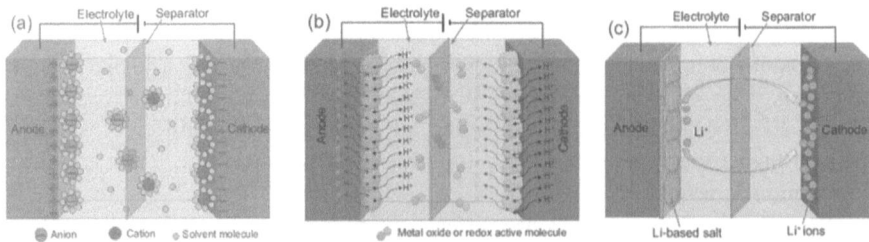

FIGURE 2.5 Schematic representation of (a) electrical double-layer capacitor (EDLC), (b) pseudocapacitor (PC), and (c) hybrid SC (HSC). (Reproduced with permission from Ref. [19] Copyright Oxford University Press (2017).)

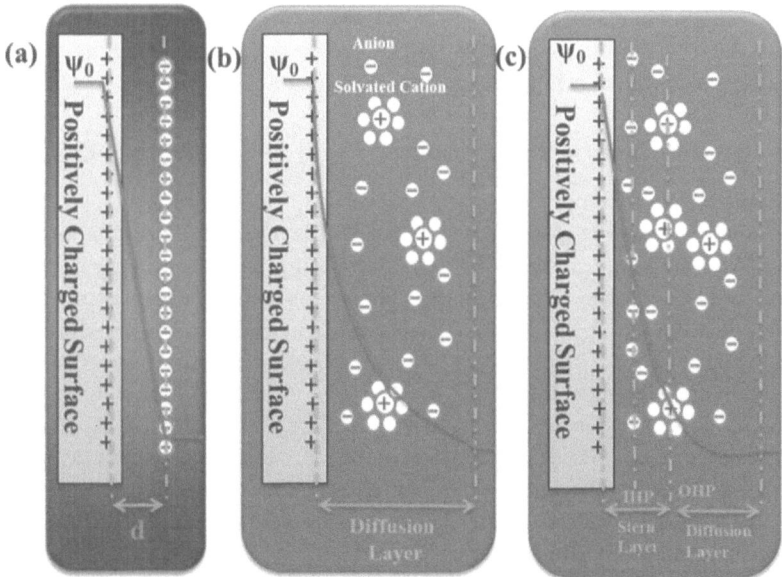

FIGURE 2.6 Models of EDL. (a) Helmholtz model. (b) Gouy-Chapman model. (c) Stern model showing the different IHP and OHP.

2.3.2 HELMHOLTZ MODEL (1853)

This is a simple theory used to model spatial load distribution. The concept of EDL was first discussed by von Helmholtz in the 19th century, as shown in Figure 2.6a and b [20]. This type of model states that the two charged layers are accumulated at a certain atomic distance on the electrode-electrolyte interface. Charge neutralization occurs through ions of opposite polarity at a distance of "d" from the solid [20–22]. The distance "d" is called the Helmholtz distance, which is the distance between the center of the ion and the charged surface. Moreover, this model should be uncertain as far as the explanation is concerned.

2.3.3 GOUY (1910)—CHAPMAN OR DIFFUSE MODEL (1913)

The simple Helmholtz model was replaced by Guy and Chapman [23,24]. They further study the distribution of electrolytes according to the thermal conductivity of electrolyte solutions. This compact layer is called the diffusion layer (Figure 2.6b). They propose a distribution similar to the opposite ion charge distribution relative to the proposed solid medium liquid, which describes ions that are not attached to the surface.

The distribution of ions in the solution is limited by the ionic meter potential of efficiency [25]. The thickness of the diffusion layer is determined in part by the ionic motive force. Furthermore, this model leads to a re-analysis of the EDL force, since the demonstrated force of the two separate loads is inversely proportional to the distance between them. In this case, when the point is charged close to the electrode surface, a higher transfer value is obtained. As already mentioned, the calibrated value of the measured thickness was lower than the test result. Therefore, this fully loaded two-layer model is not feasible [25].

2.3.4 Stern Model or Stern Modification of Diffuse Layer (1924)

The Stern model combines the models proposed by Helmholtz and Guy-Chapman. Stern found two distinct plateaus of ion distribution - the inner surface is called a dense or heavy layer (e.g., the Helmholtz layer), while the outer region is known as a sparse layer (e.g., Gouy-Chapman). The Gouy-Chapman model was further modified to determine that the ions had a finite solid size, limiting their proximity to the surface (Figure 2.6c). The properties of compact (often hydrated) layers are largely absorbed by the electrode; hence the layer is the so-called. IHP (Helmholtz Internal Aircraft) and OHP (Helmholtz External Aircraft) are used to identify the diversity of different types of ads. Besides, in addition to hydrated ions, the compact layer retains only the ads adsorbed ions and pays for non-absorbed ions. The strength of EDL (C_{dl}) can be the sum of the forces of two regions, such as the force of the diffusion region and the duplicate of the compact layer (C_H):

$$\frac{1}{C_{dl}} = \frac{1}{C_{diff}} + \frac{1}{C_H} \tag{2.1}$$

Here,

C_{dl} is the EDL capacitance

C_{diff} is the diffusion region capacitance

C_H is the double layer stern type compact layer capacitance

In SC, the electrode is usually made of a porous and high surface material area. The behavior of the EDL across the electrode pore is more complex than that of an infinitely flat surface. If the size of the pore affects the mobility of the pore ions because the pores are so small, it is not possible to influence the bilayer force capacity in this case [28]. The ED type electrode Cs (F/g) appears to be a parallel plate capacitor (by Frackowiak in 2007) [29], as follows:

$$C = A\frac{\varepsilon_r\varepsilon_0}{d} \tag{2.2}$$

where

ε_r electrolyte dielectric constant

ε_r vacuum permittivity (F/m)

A Specific surface area of ions in the electrolyte (m^2/g)

d Effective thickness of EDL (m)

According to the equation above, the relationship between force and specific surface area is proportional. Moreover, it is not suitable for experiments in many cases [30–32]. Recent studies have shown that pores smaller than 0.5 nm can easily reach hydrated ions [32,33]. Moreover, pore sizes smaller than 1 nm are often too small to obtain organic electrolytes [34]. EDL can form in micropores when hydration ions are readily soluble.

Gogotsi and his colleagues noticed a significant improvement in the power of carbon electrodes. Subsequent studies have suggested a maximum amount of energy when the size of the electrode pore is comparable to the size of ions. Therefore, these observations confirm that the size of the pore capacity is less than the contribution of the solvate ion size. These discussions and conclusions are less relevant because both scattered and compact layers are not easily placed in such confined microporous spaces.

Additionally, small mesopores on the surface of micropores contribute to improving the ability of the ultracapacitor to maintain strength. To illustrate nanoporous CO_2 emissions in 2008, Huang and colleagues proposed a heuristic approach [35,36]. Electrical mechanisms of electrodes designed with different pore sizes and the curvature of the pores are studied.

2.3.5 ELECTRIC DOUBLE-CYLINDER CAPACITOR (EDCC) MODEL

This capacitor model has been employed for mesoporous carbon electrodes along with the pore of cylindrical structures. If the pore sizes are large enough so that the pore curvature is not significant, then the EDCC model returns to the traditional planar model. The capacitance can be defined as [37]

$$C = A \frac{\varepsilon_r \varepsilon_0}{b \ln \left(\dfrac{b}{b-d} \right)} \tag{2.3}$$

2.3.6 ELECTRIC WIRE-IN-CYLINDRICAL CAPACITOR (EWCC) MODEL

Carbon electrodes of the microporous structure were modeled using this model. The ions are in a microporous position in the center of the cylindrical pore so that the capacitance is given as follows:

$$C = A \frac{\varepsilon_r \varepsilon_0}{b \ln \left(\dfrac{b}{a_0} \right)} \tag{2.4}$$

Here,
a_0 is the effective ion size
b is the radius of the pore
d distance between the carbon electrode and the approaching ion

The dielectric constant can be calculated using Equation 2.4 of the corresponding results, which were nearer to the vacuumed valve. This showed that hydrated ions dissolve completely before being introduced into micropores. Because in a more realistic environment the shape of the pores and carbon is considered rather cylindrical; another sandwich model was obtained, which is presented in the following form [38]:

$$\frac{C}{2A} = \frac{C_s}{A} \frac{\varepsilon_r \varepsilon_0}{b - a_0} \tag{2.5}$$

2.3.7 PSEUDOCAPACITORS

Capacitors of this type have disappeared, including redox reactions (charge transfer phenomena) shown in Figure 2.5b. They are made using various methods including electrospinning, redox, and intercalation. This Faraday phenomenon results in pseudocapacitors with higher Es than EDLC. Many electrode materials of this type contain metal oxides, in addition to the metal-carbon and conductive polymers [39]. However, the life of pseudo-capacitors is shorter in ps. The results are due to redox reactions in the condensers [40].

2.3.8 HYBRID SCs

Hybrid SCs have both polarizable electrodes (carbon electrodes) and non-polarizable electrodes (e.g., metal-type electrodes or polymer-type wires) as shown in Figure 2.5c. Utilizing vanishing and non-blinding properties [41], these properties are used for higher energy storage of both battery as well as capacitor electrodes [42], to provide excellent cyclic stability and reduced costs. compared to EDLC. Their three main categories are asymmetric, composite and batteries.

2.3.9 ASYMMETRIC HYBRID SCs

These capacitors are unique because of the two different electrodes. They are made to work simultaneously to meet electrical and Ed requirements, as one acts as a cathode electrode and the other as an anode electrode. In most cases, carbon-derived materials are used as negative electrodes, and a metal electrode or metal oxide is used as an anode. Metal electrodes have high internal capacity; resulting in increased energy force [43]. These capacitors have the potential to show a better Ed cycle and are also stable compared to symmetric SCs. It can be observed on carbon and MnO_2 on the surface of the nickel foam [44]. Capacitor self-discharge is a major problem for all capacitors. One way to achieve this is to use a simple mechanism and an asymmetric capacitor. The maximum potential with zero current is guaranteed here [43]. The next important task is the operating voltage of the SCs. Assuming that the selected negative electrode is powered by carbon or the sample derived from carbon is effectively doped, this could, in conjunction with the main mechanism, improve the window potential. Microporous carbon control is also a critical property that needs to be clarified. After adjustment, easy carrying of ions allows increasing the power of the device [43].

2.3.10 COMPOSITE HYBRID SCs

The main application of hybrid SCs blends is precise power to achieve synergistic results, good bike stability and higher conductivity. As can be seen from the EDLCs discussed above, carbon-based carbon has a good surface, no July heating, less strength and high mechanical strength. However, carbon itself has a low Ed level compared to the commercially charged lithium-ion and lead-acid batteries, while the metal oxide studied was poorly conductive, except for RuO_2 heating and joules and less structural stress. However, they can store charge and thus energy. capacitors of this type combine the properties of carbon oxides and metals and combine the desired synergistic properties with specific strength, good cycling stability, and greater conductivity. Together, carbon leaves the transport channel and metal oxide holds the load through redox reactions, leading to higher Cs and higher Es. Although the conductivity of the mixture is highly regulated, it is based on the structure of carbon, whether it is microporous, mesoporous, and macroporous [45], which describes that the diameter of the pore is an important factor to consider because it determines whether ions are adsorbed on the electrode surface. Functionality and download of EDLC functionality. If this is not the case then maybe there is a problem. The mixture has barriers, for example when vanadium oxide is applied to carbon nanofiber, the efficiency starts to increase because the layer is thicker (18 nm) [46]. This is due to an imbalance between the redox site and the management of the connection itself. The second problem arises when the successful diffusion of ions decreases due to the known nanofibers growing in carbon nanofibers, even though the metal oxide surface increases [47]. It shows how the limitations and advantages of the compositions depend on the components being strong, depending on their combination and the electrolytes do not exceed the scope of this study.

2.3.11   RECHARGEABLE BATTERY-TYPE HYBRID SCs

The view of this category of SC is to fight through the central diagonal of the Ragone Framework, a promising feature of higher Cs, higher Es, and advanced Ps, which however should be efficient compared to SCs. currently in use. The various factors discussed here include surface modification, precision nanocomposite material production, and microstructure optimization. Because the formation of electroactive nanoparticles causes a faster electrolyte reaction, it should effectively lead to a faster reaction, performing redox with electroactive nanoparticles.

Furthermore, it is a complex problem due to the effects of electrolyte, such as a possible wrong reaction. Several problems have arisen in the design of nanocomposite materials with metal oxides. An example of such a problem is $LiMnPO_4$, which has a higher potential than its Fe analog, but again it is difficult to cover a carbon layer similar to $LiFePO_4$. However, a good strategy approach here is to help the process with the help of Fe and Mn multilayer coal structure. Furthermore, the study says that it is better than at high speed and without the direct coupling of the oxidizing metal oxide Mn to the electrolyte. Technology Improvement - Formation of disproportionate granules and fractions on the electrode surface. Thus, the "fracture granularity" in the electrolyte-electrode interface has a larger surface effect, which increases the probability of increasing the total energy of the capacitor. Another important development is the application of a two-tiered approach to this type of SC. Create a Helmholtz double layer where the charge is held at a carbon electrode and an electrolyte interface. This is since such loads differ from each other in the interface, counteracts, accelerates the mechanism of accumulation of physical load due to the formation of polar ions. This phenomenon is due to the effect of the change of SC Es.

2.4 INTEGRATION OF SUPERCAPACITORS TO GRID

Disruptions or disturbances may occur at any time without warning. Some of the largest companies in the world have suffered losses of up to hundreds of millions of dollars due to disruptions that lasted seconds or even milliseconds. Battery-based UPS solutions are used everywhere, but sometimes they do not respond quickly or fail due to a faulty system.

Ultracapacitors are an ideal solution in case of a short electrical fault, as they are both highly reliable and can react quickly. The growing share of renewable energies increases grid uncertainty. The global grid and renewable energy markets are expanding rapidly. The growing volume of renewable energy sources, such as solar energy and wind energy, requires energy storage technology to balance the energy load and reduce the mains voltage. Ultracapacitors help prevent interruptions by providing maximum power and extending battery life in energy storage. Ultracapacitors can also back up air control systems without interrupting the power supply (UPS), as well as other short-term and modern applications (Figure 2.7).

The ultracapacitor-based energy storage provides solutions either by supplying ultracapacitor modules to be integrated into a larger system or to be delivered as a large turn-key solution in partnership with System Integrators (Figure 2.8).

The grid systems can reach high levels of complexity and become quite daunting to control. Within them, there can be an array of different applications with various means of energy storage associated with them. Ultra-capacitors as an energy storage medium can provide big benefits such as instant response to demand surges and improved energy delivery quality.

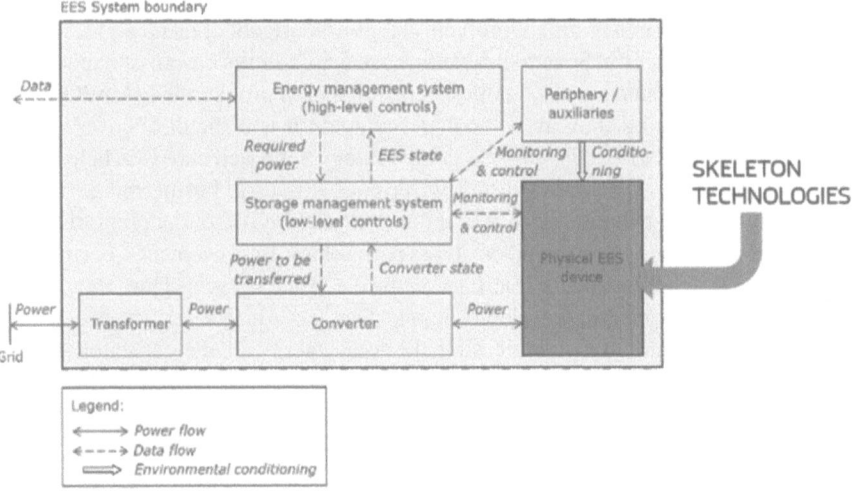

FIGURE 2.7 Schematic for power and data flow for the EES system boundary.

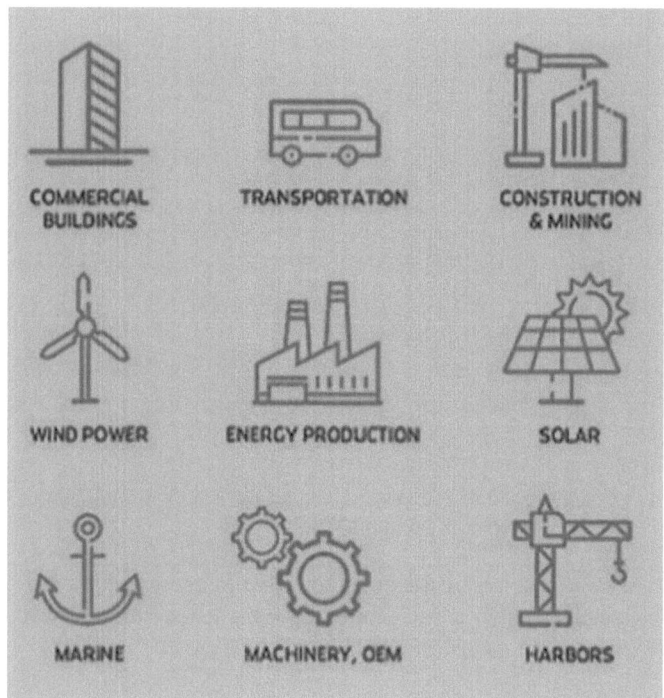

FIGURE 2.8 Skeleton's competence in grid & behind meter applications.

2.4.1 POWER QUALITY AND VOLTAGE DISTURBANCES

Ultracapacitor energy storage can be an ideal solution for a variety of energy quality problems, including coverage, glare, sinking, swelling, and interruptions, as they can respond almost instantly (Figure 2.9).

Semiconductor manufacturing, pharmaceutical industry (or any clean space environment), automotive industry, metal presses, painting, and many other industries

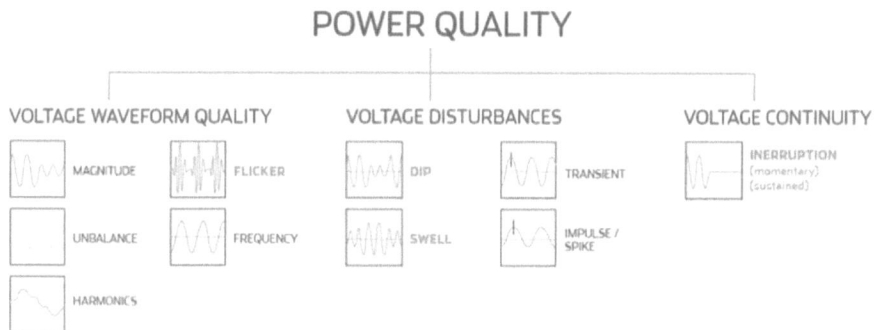

FIGURE 2.9 Skeleton technologies grid and renewable energy power quality.

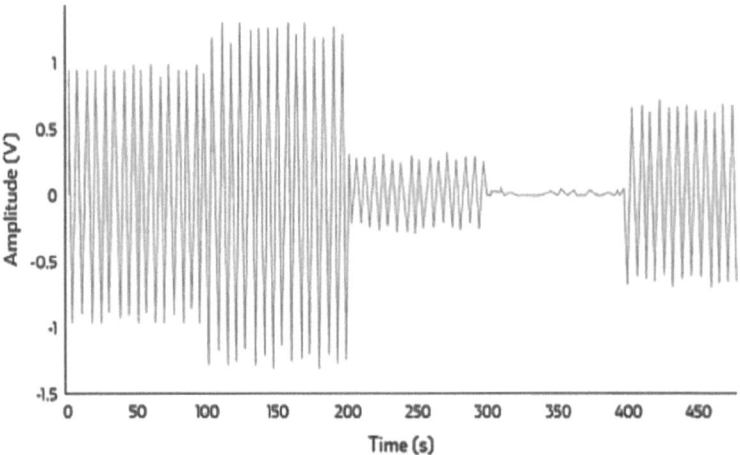

FIGURE 2.10 Skeleton technologies grid and renewable energy power quality.

and production processes are particularly sensitive to energy quality problems and dirt caused by a single quality event. Low energy can be measured in tens of millions of dollars (Figure 2.10).

2.4.2 SUPERCAPACITORS: INTEGRATION INTO WIND/SOLAR GENERATION SYSTEMS – IMPACT POINT

In the case of an integrated energy-efficient energy conversion system, the energy efficiency of the system and the energy efficiency of the energy supply system of the administrator must be maintained. It is possible to adjust the battery to the density of energy and, in the case of electricity, to the supply of electrical energy [48,49]. In this case, the supercapacitor can be used as a power supply, even if it is used, or if it is repeatedly discharged [50]. It is possible to provide and dispose of a hybrid battery ultracondenser, a continuation, and a fibrillation system that follows a common Eolian solar array [51].

In January 2017, the US wind energy identification capacity was 82,183 megawatts (MW) [52]. Wind energy, like other renewable energy sources, is variable and subject to fluctuations and it is therefore extremely important to store excess energy during times of strong winds and low demand and then release some of this energy [53]. Having a supercapacitor and a hybrid system with batteries and a "storage system" helps to effectively reduce short-term air energy differences. Thus, the battery cycle can be safely controlled, with the full advantage of prolonging its life [54]. Using supercapacitors in a hybrid system can reduce the number of battery cycles and improve their life. Unlike batteries, supercapacitors can withstand a greater number of cycles, which is a weak link in the system [55]. Wind energy can also damage the stability of the mains voltage. Supercapacitors and a built-in hybrid storage system can provide sufficient reactive energy to maintain local voltage levels [56].

All independent photovoltaic systems require an energy buffer to cover the gap between production and energy consumption. Battery technology, mainly lead-acid batteries, is the most popular form of photovoltaic energy storage. However, batteries are not ideal storage systems for photovoltaic cells due to their unreliable weather conditions. Inadequate battery charging usually leads to sulfate or coagulation, both of which shorten the battery life. "Search" also reduces battery life due to battery overcharging. Supercapacitors can be integrated into batteries in a hybrid photovoltaic system. The hybrid system offers the benefits of both technologies, in both high energy density and energy density. A supercapacitor in a hybrid system is used to charge the battery when the panel delivers excess power and is an excellent choice for high-end applications due to its high power density [57]. In a hybrid system, the supercapacitor runs on battery power to deliver load power when the solar panel does not produce enough energy. In this situation, the battery offers a low constant load, while the supercapacitor guarantees maximum power consumption. This approach results in higher battery life and reduced battery size. Of course, there are many advantages to using a hybrid battery supercapacitor system as a solar and wind power storage system. A hybrid system can stabilize the power supply not only for repetitive energy but also for unstable loads [58].

2.5 CONCLUSIONS

The various classifications and operating principles of the supercapacitors reviewed in this chapter are diligently descriptive. False properties and internal mechanisms of these reactions were recorded with great interest. The importance of supercapacitors in grid integration plays a key role in effectively supporting the battery energy storage system and increasing the power at different operating levels. The rapid growth of renewable energy sources requires innovation and testing of new storage technologies, such as the supercapacitor battery hybrid system. This chapter does the same and provides a thorough overview of the use of the Hybrid Supercapacitor Renewable Battery System. The combination makes energy storage at both high energy and energy density, which extends battery life.

REFERENCES

1. L.L. Zhang, X.S. Zhao, Carbon-based materials as SC electrodes, *Chem. Soc. Rev.* 38 (2009), 2520–2531.
2. A. Balducci, R. Dugas, P.L. Taberna, P. Simon, D. Plée, M. Mastragostino, S. Passerini, High-temperature carbon-carbon SC using ionic liquid as electrolyte, *J. Power Sources* 165 (2007), 922–927.
3. C. Largeot, C. Portet, J. Chmiola, P.L. Taberna, Y. Gogotsi, P. Simon, Relation between the ion size and pore size for an electric double-layer capacitor, *J. Am. Chem. Soc.* 130(9) (2008), 2730–2731.
4. S.G. Kandalkar, D.S. Dhawale, C.K. Kim, C.D. Lokhande, Chemical synthesis of cobalt oxide thin film electrode for SC application, *Synth. Met.* 160 (2010), 1299–1302.
5. M. Winter, R.J. Brodd, What are batteries, fuel cells, and SCs? *Chem. Rev.* 104 (2004), 4245–4269.
6. B.E. Conway, Transition from "SC" to "Battery" behavior in electrochemical energy storage, *J. Electrochem. Soc.* 138(6) (1991), 1539–1548.

7. Z. Mingjia, X. Chengcheng, L. Jiaangtian, L. Ming, W. Nianqiang, Nanostructured carbon-metal oxide composite electrodes for SCs: a review, *Nanoscale* 5 (2013), 72–88.

8. S. Wu, Y. Zheng, S. Zheng, S. Wang, C. Sun, K. Parvez, T. Ikeda, X. Bao, K. Mullen, X. Feng, Stacked-layer heterostructure films of 2D thiophene nanosheets and graphene for high-rate all-solid-state pseudocapacitors with enhanced volumetric capacitance, *Adv. Mater.* 29 (2017), 1602960.

9. S. Kondrat, A.A. Kornyshev, Pressing a spring: what does it take to maximize the energy storage in nanoporous SCs? *Nanoscale Horiz.* 1 (2016), 45–52.

10. J. Tang, Y. Yamauchi, Carbon materials: MOF morphologies in control, *Nat. Chem.* 8 (2016), 638–639.

11. S. Zheng, Z.S. Wu, S. Wang, H. Xiao, F. Zhou, C. Sun, X. Bao, H.M. Cheng, Graphene-based materials for high-voltage and high-energy asymmetric SCs, *Energy Storage Mater.* 6 (2017) 70–97.

12. G. Wang, L. Zhang, J. Zhang, A review of electrode materials for electrochemical SCs, *Chem. Soc. Rev.* 41 (2012), 797–828.

13. https://www.maximizemarketresearch.com/market-report/global-SC-market/27055/.

14. https://www.thomasnet.com/articles/top-suppliers/capacitor-manufacturers-suppliers/.

15. R.B. Ambade, S.B. Ambade, N.K. Shrestha, R.R. Salunkhe, W. Lee, S.S. Bagde, J.H. Kim, F.J. Stadler, Y. Yamauchi, S.H. Lee, Controlled growth of polythiophene nanofibers in TiO_2 nanotube arrays for SC applications, *J. Mater. Chem. A* 5 (2017), 172.

16. Hassan M. Reddy K.R., Haque E., Faisal S.N., Ghasemi S. Hierarchical assembly of graphene/polyaniline nanostructures to synthesize free-standing SC electrode. *Compos. Sci. Technol.* 98 (2014), 1–8.

17. K.R. Reddy, B.C. Sin, C.H. Yoo, W. Park, K.S. Ryu, J.S. Lee, D. Sohn, Y. Lee, A new one-step synthesis method for coating multi-walled carbon nanotubes with cuprous oxide nanoparticles, *Scr. Mater.* 58 (2008), 1010–1013.

18. K.R. Reddy, B.C. Sin, K.S. Ryu, J. Noh, Y. Lee, In situ self-organization of carbon black-polyaniline composites from nanospheres to nanorods: synthesis, morphology, structure and electrical conductivity, *Synth. Met.* 159 (2009), 1934–1939.

19. X. Chen, R. Paul, L. Dai, Carbon-based SCs for efficient energy storage, *Natl. Sci. Rev.* 4 (2017), 453–489.

20. L.L. Zhang, X.S. Zhao, Carbon-based materials as SC electrodes, *Chem. Soc. Rev.* 38 (2009), 2520–2531

21. X. Xiao, G. Wang, M. Zhang, Z. Wang, R. Zhao, Y. Wang, Electrochemical performance of mesoporous $ZnCo_2O_4$ nanosheets as an electrode material for SC, *Ionics (Kiel).* 24 (2018), 2435.

22. H. Zhang, X. Lv, Y. Li, Y. Wang, J. Li, P25-Graphene composite as a high performance photo catalyst. *ACS Nano* 4 (2009), 380–386.

23. G. Gouy, A Modified Theory of the Electrical Double Layer. *J. Phys. Theor. Appl.*, 9(1) (1910), 457–468.

24. D.L. Chapman, LI. A contribution to the theory of electrocapillarity. *Philos. Mag.* 6 (1913), 475–481.

25. M. Endo, T. Takeda, Y. Kim, K. Koshiba, K. Ishii, High power electric double layer capacitor (EDLC's); from operating principle to pore size control in advanced activated carbons. *CarbonScience* 1(3) (2001), 117–128.

26. O. Z. Stern, Theory of the electrolytic double layer, *Elek- trochem.* Angew. Phys. Chem., 30 (1924), 508–516.

27. L.L. Zhang, X.S. Zhao, Carbon-based materials as SC electrodes, *Chem. Soc. Rev.* 38(9) (2009), 2520–2531.

28. P. Sharma, T.S. Bhatti, A review on electrochemical double-layer capacitors, *Energy Convers. Manag.* 51(12) (2010), 2901–2912.

29. O. Barbieri, M. Hahn, A. Herzog, R. Kötz, Capacitance limits of high surface area activated carbons for double layer capacitors, *Carbon N. Y.* 43(6) (2005), 1303–1310.
30. D. Qu, H. Shi, Studies of activated carbons used in double-layer capacitors, *J. Power Sources* 74(1) (1998), 99–107.
31. J. Gamby, P. Taberna, P. Simon, J. Fauvarque, M. Chesneau, Studies and characterisations of various activated carbons used for carbon/carbon SCs, *J. Power Sources* 101(1) (2001), 109–116.
32. H. Shi, Activated carbons and double layer capacitance, *Electrochim. Acta* 41(10) (1996), 1633–1639.
33. D. Qu, Studies of the activated carbons used in double-layer SCs, *J. Power Sources* 109(2) (2002), 403–411.
34. Y. Kim, Y. Horie, S. Ozaki, Y. Matsuzawa, H. Suezaki, C. Kim et al Correlation between the pore and solvated ion size on capacitance uptake of PVDC-based carbons, *Carbon* 42(8–9) (2004), 1491–1500.
35. J. Huang, B.G. Sumpter, V. Meunier, Theoretical model for nanoporous carbon SCs, *Angew. Chem. Int. Ed.* 47 (2008), 520–524.
36. J. Huang, B.G. Sumpter, V. Meunier, A universal model for nanoporous carbon SCs applicable to diverse pore regimes, carbon materials, and electrolytes, *Chem. Eur. J.* 14 (2008b), 6614–6626.
37. J. Huang, B. Sumpter, V. Meunier, Theoretical model fornanoporous carbon SCs, *Angew. Chem.* 120(3) (2008), 530–534.
38. G. Feng, R. Qiao, J. Huang, B.G. Sumpter, V. Meunier, Ion distribution in electrified micropores and its role in the anomalous enhancement of capacitance, *ACS Nano* 4(4) (2010), 2382–2390.
39. M. Vangari, T. Pryor, L. Jiang, SCs: review of materials and fabrication methods, *J. Energy Eng.* 139 (2013), 72–79.
40. M.D. Stoller, S. Park, Z. Yanwu, J. An, R.S. Ruoff, Graphene-based ultracapacitors, *Nano Lett.* 8 (2008), 3498–3502.
41. J.C.E. Halper, M.S. Halper, SCs: A Brief Overview, MITRE Nanosystems Group, March 2006.
42. M.Y. Ho, P.S. Khiew, D. Isa, T.K. Tan, W.S. Chiu, C.H. Chia, A review of metal oxide composite electrode materials for electrochemical capacitors, *Nano* 9 (2014), 1430002.
43. H.D. Yoo, S.-D. Han, R.D. Bayliss, A.A. Gewirth, B. Genorio, N.N. Rajput, K.A. Persson, A.K. Burrell, J. Cabana, "Rocking-Chair"-type metal hybrid SCs, *ACS Appl. Mater. Interfaces*, 08 (2016), 30853–30862.
44. H. Gao, F. Xiao, C.B. Ching, H. Duan, High-performance asymmetric SC based on graphene hydrogel and nanostructured MnO_2, *ACS Appl. Mater. Interfaces* 4(5) (2012), 2801.
45. M. Zhi, C. Xiang, J. Li, M. Li, N. Wu, Nanostructured carbon-metal oxide composite electrodes for SCs: a review, *Nanoscale* 5 (2013), 72–88.
46. A. Ghosh, E.J. Ra, M. Jin, H.K. Jeong, T.H. Kim, C. Biswas, Y.H. Lee, High pseudocapacitance from ultrathin V_2O_5 films electrodeposited on self-standing carbon-nanofiber paper, *Adv. Funct. Mater.* 21 (2011), 2541–2547.
47. M. Zhi, A. Manivannan, F. Meng, N. Wu, Highly conductive electrospun carbon nanofiber/MnO_2 coaxial nano-cables for high energy and specific power SCs, *J. Power Sources* 208 (2012), 345–353.
48. https://www.iea.org/topics/renewables/.
49. http://berc.berkeley.edu/storage-wars-batteries-vs-supercapacitors/.
50. W. Hongbin, C. Bin, G. Caiyun, Capacity optimization of hybrid energy storage units in wind/solar generation system, *Trans. Chinese Soc. Agricult. Eng.* 27(5) (2011), 241–245.
51. https://www.awea.org/MediaCenter/pressrelease.aspx?ItemNumber=9812.

52. M. Black, G. Strbac, Value of bulk energy storage for managing wind power fluctuations, *IEEE Trans. Energy Conversion* 22(1) (2007), 197–205.

53. M.A. Tankari, M.B. Camara, B. Dakyo, G. Lefebvre, Use of ultracapacitors and batteries for efficient energy management in wind–diesel hybrid system, *IEEE Trans. Sustain. Ener.* 4(2) (2013), 414–424.

54. A. Khaligh, Z. Li, Battery ultracapacitor fuel cell and hybrid energy storage systems for electric hybrid electric fuel cell and plug-in hybrid electric vehicles: state of the art, *IEEE Trans. Veh. Technol.* 59(6) (2010), 2806–2814.

55. H. Zhao, Q. Wu, S. Hu, H. Xu, C.N. Rasmussen, Review of energy storage system for wind power integration support, *Appl. Ener.* 137 (2015), 545–553.

56. M.E. Glavin, W.G. Hurley, Ultracapacitor/battery hybrid for solar energy storage. *Proceedings of the 42nd International Universities Power Engineering Conference*, 2007. UPEC 2007.

57. *Proceedings of the 16th International Seminar on Double Layer Capacitors and Hybrid Energy Storage Devices*, December 2006.

58. Ma, Tao & Yang, Hongxing & Lu, Lin, Development of hybrid battery -supercapacitor energy storage for remote area renewable energy systems. *Applied Energy*, Elsevier, 153(C) (2015), 56–62.

3 Implementation of Solar, UC/Battery-Based Hybrid Electric Vehicle with an Efficient Controller

Geetha Reddy Evuri
Teegala Krishna Reddy College of Engineering
and Technology (TKRCET)

M. Divya
Vignan's Lara Institute of Technology & Science

K. Srinivasa Reddy
Teegala Krishna Reddy Engineering College

B. Nagi Reddy
Vignana Bharathi Institute of Technology (VBIT)

G. Srinivasa Rao
Vignan's Foundation for Science, Technology and Research

CONTENTS

DOI: 10.1201/9781003143802-3

3.1   INTRODUCTION

Owing to the increasing custom of fuel-based vehicles, universal heating has been increasing over the last few years. To reduce global warming and pollution, a solution for the conventional vehicles has to be designed. It leads to the utilization of renewable energy sources avoiding the pollution. The hybrid electric vehicle [1–3] which runs on battery, ultra capacitor (UC) and solar panels, has been chosen, which reduces pollution and provides an alternative for the future generation. Solar energy has been chosen as the energy source among a number of renewable energy sources due to its advantages of being non pollutant and unlimited and so the electric drive vehicle or hybrid vehicle considered in this research study is a solar powered vehicle. The power from the solar energy is not constant and it varies according to the irradiance, so as to obtain the maximum power irrespective of the weather conditions an (Maximum Power Point Tracking) controller of which P&O has been used for the solar panel. A two input full bridge buck boost converter has been selected as a DC–DC converter [5–7] among a number of conventional DC–DC converters like buck, boost converter, buck boost converter, etc. due to its advantages like utilizing the transformer efficiently, limiting the voltage stress among all the switches. The two inputs to the converter are battery and UC. The lead acid battery [8,9] has been chosen among other batteries like lithium ion and nickel metal hydride due to its advantages of being simple, inexpensive, environmentally safe, and available in high recycle range. In addition to battery, UC has been considered as another energy storage device to increase the battery life and provide energy during transient conditions. The efficiency characteristics of battery, UC and electric motor in a hybrid vehicle have been analyzed with an optimal efficiency model proposed, based on interfacing super capacitors with batteries. It is required to use an energy storage device [10] for storing excess energy and can be used in case of power depletion in the production. Among a number of available batteries, the lithium ion nickel metal battery can accumulate a huge quantity of energy but the energy in these types of batteries fall during rapid change in the load. UCs [11–15] are the one which can act with the batteries and can be used to increase the life time of the batteries and also to provide excessive energy during transient period of operations. To control the output of the DC–DC converter, controller is required where PI controller has been chosen due to its zero steady state error, stability and better maximum peak overshoot compared to the individual integral controller. The output from the batteries has been fed to the brushless DC (BLDC) motors for running the vehicle. The BLDC motor has been chosen due to its advantages like higher efficiency, smaller in size, and no copper losses due to absence of filed windings.

   This chapter is organized as follows: Section 3.1 presents introduction to the proposed hybrid electric vehicle, Section 3.2 deals with the block diagram explanation of the proposed hybrid electric vehicle, Section 3.3 deals with the mathematical modeling of the proposed hybrid electric vehicle, Section 3.4 deals with the discussion of proposed controller, Section 3.5 deals with the simulation work and results. Section 3.6 deals with hardware implementation of the proposed hybrid vehicle and the comparison between the simulation and hardware results, and Section 3.7 deals with the conclusions drawn from the proposed work.

3.2 BLOCK DIAGRAM OF THE PROPOSED HYBRID VEHICLE

The block diagram for the proposed hybrid vehicle is shown in Figure 3.1. When the solar rays are injected into the solar panel, the solar energy gets converted into DC energy. The DC energy obtained from the solar panel is used to charge the battery and UC. So as to maintain constant output to the electric vehicle, two input full bridge buck boost converters have been designed with MOSFET switches which can buck or boost the output based on the requirement of the battery. The output of the DC–DC converter may not be constant due to variations with respect to the load and in order to provide the constant output, a PI controller has been designed to track the reference value of the output voltage. The constant output from the DC–DC converter is fed to the motor drive through a three-level inverter. The UC can be operated in the transient condition to support the operation of the battery. The three-level inverter converts the output DC value from the converter into three phase three-level AC which is fed to the BLDC motor for driving the vehicle.

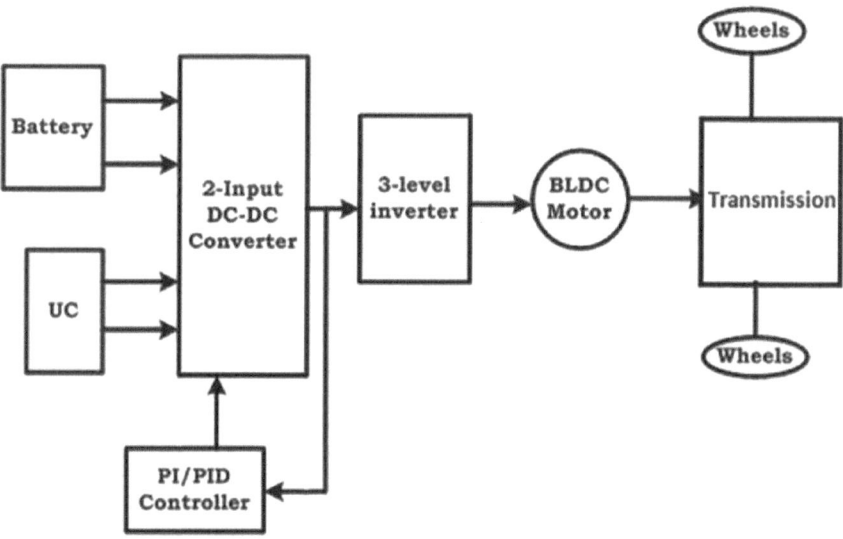

FIGURE 3.1   Block diagram of the proposed hybrid electric vehicle.

3.3 MATHEMATICAL MODELING OF THE PROPOSED HYBRID VEHICLE

3.3.1 SOLAR PANEL

The solar panel contains multiple solar cells connected in series and parallel to obtain the required output whose equivalent model is shown in Figure 3.2.

The equation for the current coming out from the solar cell can be given as

$$I = I_{ph} - I_D - I_{sh} \tag{3.1}$$

Where the current flowing through the diode can be given as

$$I_D = I_0 \left[e^{q\left(\frac{V + IR_S}{AKT}\right)} - 1 \right] \tag{3.2}$$

The current flowing through the shunt resistance can be given as

$$I_{sh} = \frac{V + IR_S}{R_{sh}} \tag{3.3}$$

3.3.2 FULL BRIDGE DC–DC CONVERTER

Due to some of the disadvantages of conventional buck boost converter [13] like high stress, an isolated two input full bridge converter has been proposed in this chapter whose circuit diagram is shown in Figure 3.3. It consists of two full bridge converters whose operation is described below. It is a buck boost converter which can be used in medium and high power applications. It consists of eight switches, transformer and four diodes for converting AC obtained from the transformer into DC supply.

The function of the converter has been analyzed into four modes which is explained below.

> **Mode I**: In this approach of operation the two switches, Q1, Q2 are switched ON and the power is transferred from input to the output and the power is converted into DC power using one pair of diodes in the rectifier circuit.
>
> **Mode II**: In this form of operation, the switching device Q1 is switched OFF and the capacitor gets the energy from the transformer windings through the pair of diodes.

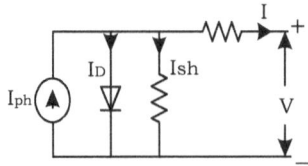

FIGURE 3.2 Equivalent circuit of solar cell.

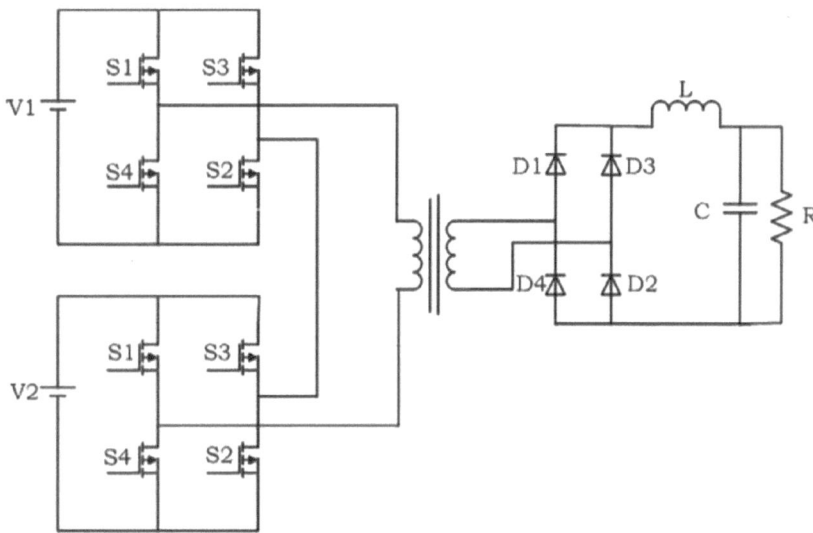

FIGURE 3.3 Circuit diagram of two input full bridge DC–DC converter and its waveforms.

Mode III: In this approach of operation, the switch Q4 is switched ON and the capacitor gets the power starting the transformer windings through the pair of diodes.

Mode IV: In this type of operation, the switching device Q2 is turned OFF, then after an interval of time, the device Q3 is turned ON. In this way the cycle repeats as well as the converted energy can be obtained whose waveform is shown in the below figure.

The equation for the voltage at the rectifier can be given as

$$V_{rec} = DV_{in}n \tag{3.4}$$

The equation for the current at the input can be given as

$$I_{in} = DI_L n \tag{3.5}$$

3.3.3 BATTERY

A battery is an apparatus which converts the chemical power into electrical power and electrical power into chemical power. It contains figure of voltaic cells where each voltaic cell consists of two cells associated in sequence with an electrolyte. Owing to the chemical actions within these cells, the ability of the battery depends upon the enormity of current, duration, terminal voltage and temperature. The existing capacity of the battery depends upon the velocity at which it is discharged and if the battery discharges at high pace, the ability decreases at a faster rate than expected (Figure 3.4).

The relation between the charge of the battery and current through the battery can be related as

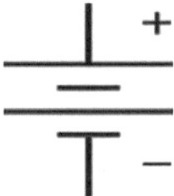

FIGURE 3.4   Symbol for a battery.

$$I = \frac{Q}{t} \tag{3.6}$$

Where Q is the battery capacity, I is the current flowing through the battery and t is the time phase that a battery can maintain.

3.3.4   ULTRA CAPACITOR

An UC also well-known as dual film capacitor provides an electrolytic resolution for storing the power. Even though it is an electrochemical device, no chemical reactions are implicated in its power storage principle and storage principle intends the UC to allege and eject a number of times.

An UC as shown in Figure 3.5 consists of two non reactive porous electrodes suspended within the electrolyte with the application of voltage across the collectors. In an entity cell, the functional voltage on the positive electrode attracts the negative ions and the negative electrode attracts the positive ions. Dielectric separators amid the two electrodes prevent the charge from stirring the electrodes.

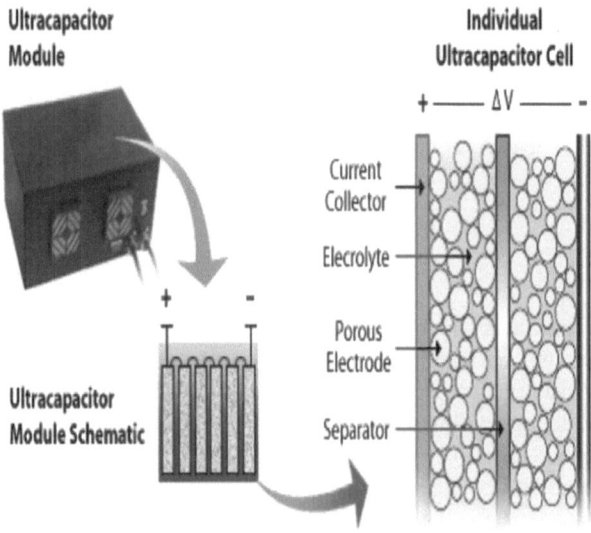

FIGURE 3.5   Diagram depicts an ultra capacitor, its modules, and an ultra capacitor cell.

Once the dual film capacitor is fully stimulated and stores the energy, the vehicle can make use of the power. The amount of energy stored in this UC is very huge in comparison to the normal capacitor because of the surface area formed by porous carbon electrodes and the charge partition formed by the dielectric barrier. But it stores the energy much less than the normal battery whereas the UC can discharge the energy much faster than the battery and provides large bursts of energy. UCs eliminate the instantaneous energy demand on the battery and extend battery runtime and also reduce declining of charge capacity of the battery. These are maintenance free and can be operated over a broad array of temperature. These UCs can be operated in conjunction with batteries or other energy sources like photovoltaic cells, wind power generation and fuel cells when the battery alone does not provide the required energy.

3.3.5 BRUSHLESS DC (BLDC) MOTOR

Conventional DC motors have the disadvantages of requiring a commutator, brushes leading to wear and tear, and sparking at the surface of the brushes. To overcome these disadvantages of the DC motor, a BLDC motor has been proposed for the hybrid vehicle. The three phase output of the inverter is fed to the BLDC motor which is used to drive the vehicle and BLDC is a normal DC motor whose coil is placed in the stator and permanet magnets are placed in the rotor (Figure 3.6).

The three phase-controlled output of the three-level SPWM inverter is fed to the BLDC motor which rotates and generates mechanical energy and three phase back EMFS. The equivalent circuit of the motor is shown in Figure 3.7 which consists of resistance and inductance and back EMF.

The dynamic equations of BLDC motor can be given as

$$V_a = i_a R_a + L_a \frac{di_a}{dt} + e_a \tag{3.7}$$

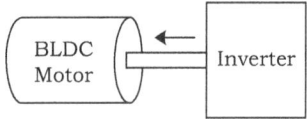

FIGURE 3.6 Block diagram of BLDC with inverter.

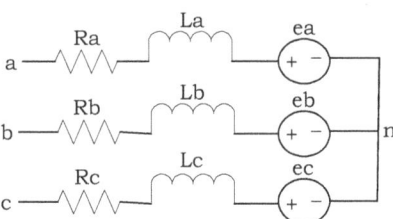

FIGURE 3.7 Equivalent circuit of BLDC motor.

$$V_b = i_b R_b + L_b \frac{di_b}{dt} + e_b \qquad (3.8)$$

$$V_c = i_c R_c + L_c \frac{di_c}{dt} + e_c \qquad (3.9)$$

The electromagnetic torque developed with in the BLDC motor can be given as

$$T_e = J \frac{dw}{dt} + Bw + T_L \qquad (3.10)$$

3.4 PROPOSED CONTROLLER

3.4.1 PI CONTROLLER

A controller is a circuit used to influence the system through a control signal so as to make the controlling variable equal to the reference value. It basically produces a value to the controlling element based on the variation among the reference value and the measured value. A proportional integral controller is the combination of proportional and integral controller (P+I) which are connected in parallel and with the proper tuning of proportional and integral gain terms (Figure 3.8).

The equation for the PI controller is given as

$$y(t) = K_p e(t) + K_i \int e(t) dt \qquad (3.11)$$

As it is arrangement of proportional and integral parts, the system becomes more steady state and less responsive to the noise signals but the system is slower.

3.4.2 PI CONTROLLER

Proportional integral derivative controller is the arrangement of proportional, integral and derivative controller (P+I+D) (Figure 3.9).

$$y(t) = K_p e(t) + K_i \int e(t) dt + K_d \frac{de(t)}{dt} \qquad (3.12)$$

In the largest part of the PID controllers, D-control retort rely only on process variable, relatively than inaccuracy. This avoids spikes in the amount produced in crate

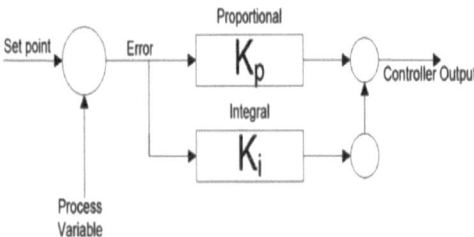

FIGURE 3.8 Block diagram of proportional integral controller.

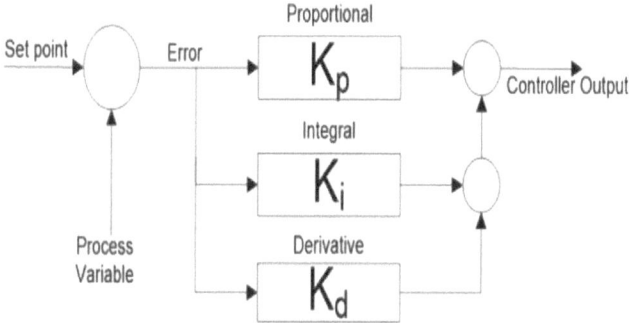

FIGURE 3.9 Proportional derivative integral controller.

of sudden set point alteration by the operator. Also, the majority control systems use a lesser amount of derivative time, t_d, as the derivative retort is very susceptible to the noise in the process changeable which leads to produce enormously high output yet for a small amount of noise.

Therefore, by the combination of proportional, integral, and derivative control outputs, a PID controller is produced. A PID controller produces a universal function; however, the PID settings should be known to adjust it properly to generate the desired output. Fine-tuning means the method of getting an ideal reaction from the PID controller by locating the most favorable gains of proportional, integral and derivative parameters.

3.5 SOFTWARE IMPLEMENTATION OF THE PROPOSED HYBRID VEHICLE

The proposed hybrid vehicle has been simulated using MATLAB simulink and the results are discussed below. Figure 3.10 shows the voltage, current characteristics and

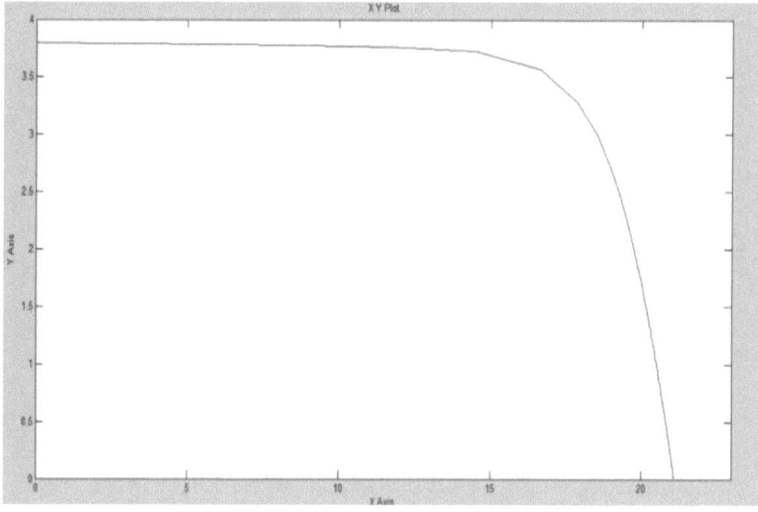

FIGURE 3.10 Voltage current characteristics of solar panels.

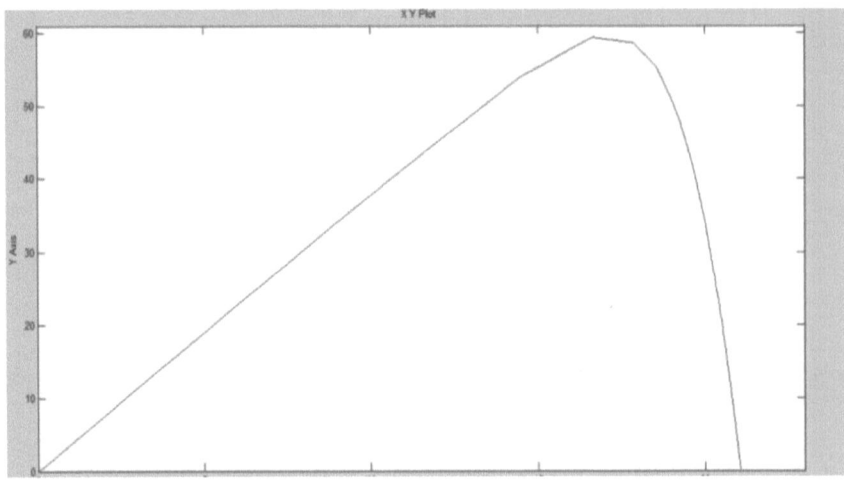

FIGURE 3.11 Power voltage characteristics of solar panels.

power, voltage characteristics obtained from the solar panel with the series connection of solar cells (Figure 3.11).

The above characteristics shows the voltage, current and power characteristics of solar panels (Figure 3.12).

The above waveform shows the voltage across UC whose value is 1.8 V (Figure 3.13).

The above waveform shows the current through the battery whose value is 11 A (Figure 3.14).

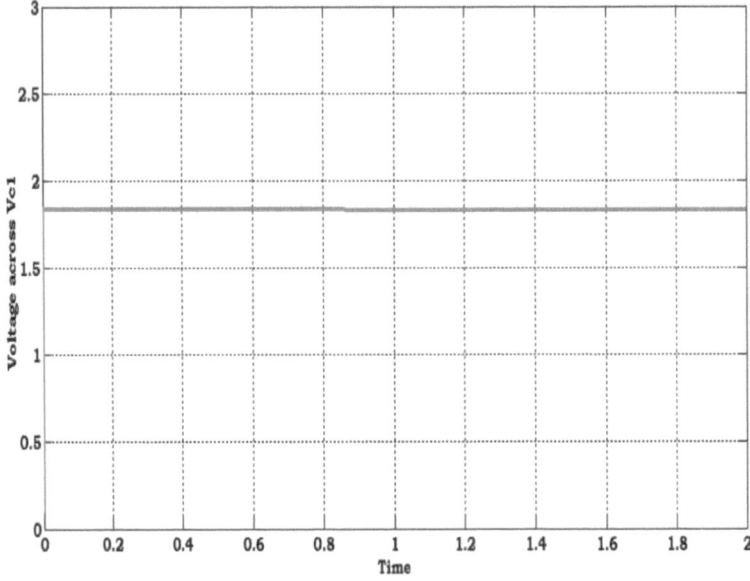

FIGURE 3.12 Voltage across UC.

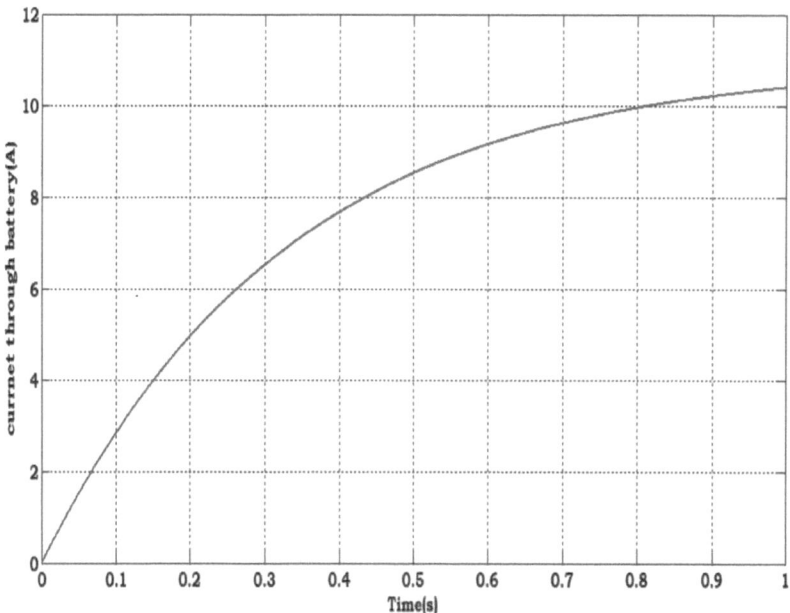

FIGURE 3.13 Current through battery.

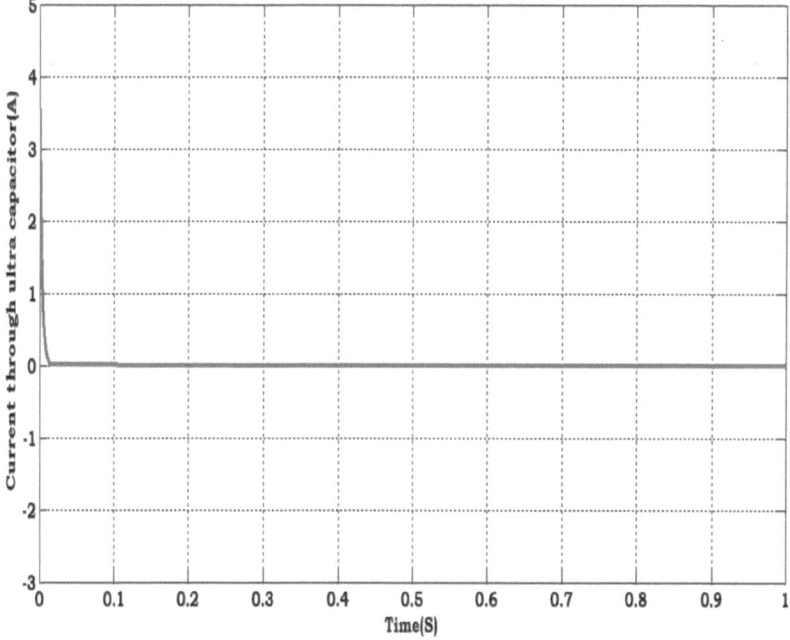

FIGURE 3.14 Current through UC.

TABLE 3.1
Comparison between PI and PID Controllers

Parameter	PI	PID
Rise time, t_r (s)	0.01	0.02
Settling time, t_s (s)	0.17	0.1
Peak time, t_p (s)	0.03	0.03
Peak overshoot (%)	10	7
Steady state error, (E_{ss})	0.7	0.6
Efficiency (%)	90.1	90.5
THD (%)	4.11	3.24

The output of the solar panel cannot be directly used for operating the vehicle as its output is very low and so to boost up the voltage to the required value (Table 3.1), DC–DC converter is connected to the solar panel and the boosted voltage is shown in Figures 3.15–3.17.

The output of the converter is fed to the three phase inverter to convert the DC of the converter to the AC output and fed to the BLDC motor which is shown in Figure 3.18.

Figure 3.19 shows the torque and speed, current and EMF of the BLDC motor to operate the electric vehicle.

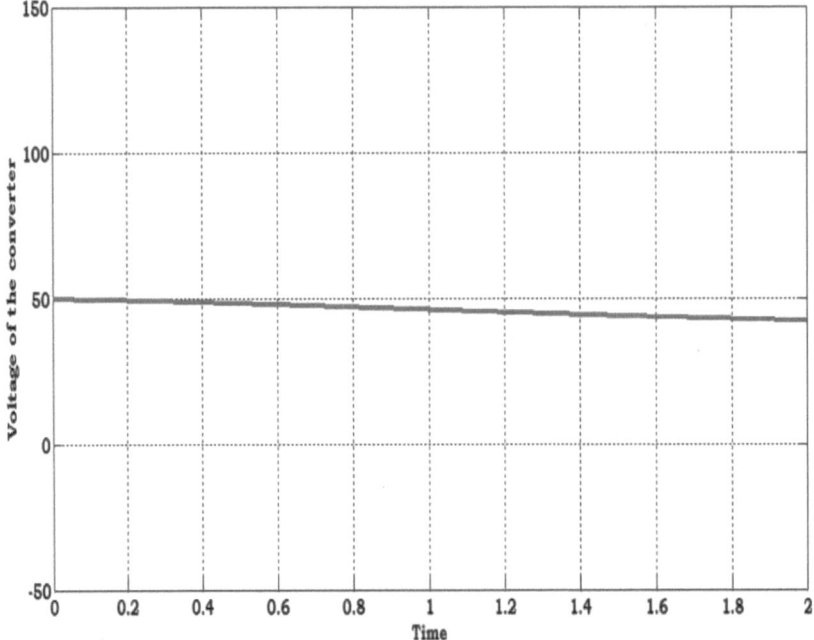

FIGURE 3.15 Output voltage of DC–DC converter.

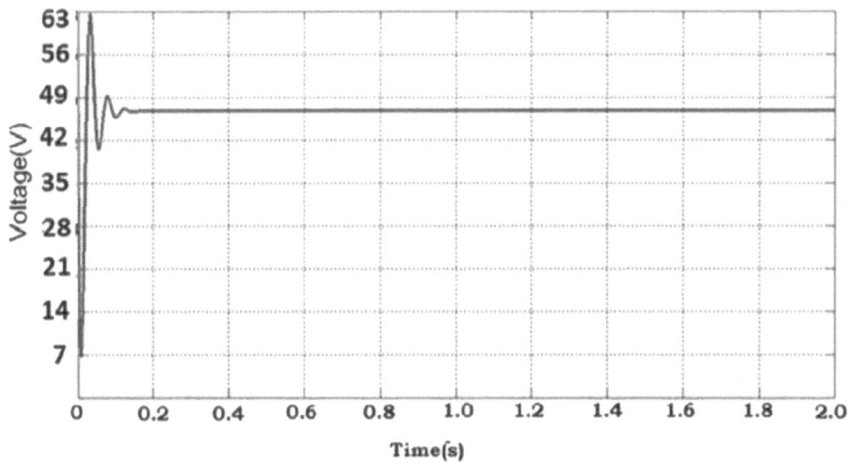

FIGURE 3.16   Output voltage of DC–DC converter with PI controller.

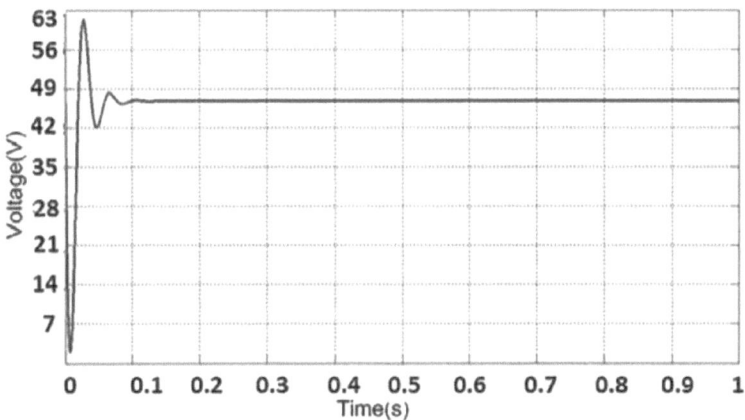

FIGURE 3.17   Output voltage of DC–DC converter with PID controller.

3.6   HARDWARE IMPLEMENTATION OF THE PROPOSED HYBRID VEHICLE

The hardware model of the proposed electric vehicle has been fabricated and the below figure shows the fabricated model and its internal structure (Figure 3.20). The proposed vehicle has been implemented in the hardware model and the flowchart for the vehicle control unit is shown in Figure 3.21.

The ports will be initialized and then switch status will be taken as the input. If the status of the switch is high, then acceleration will occur in the forward direction; if the status of the switch is low, then acceleration will occur in the reverse direction. The flowchart for the vehicle capacitor bank control unit is shown in Figure 3.22.

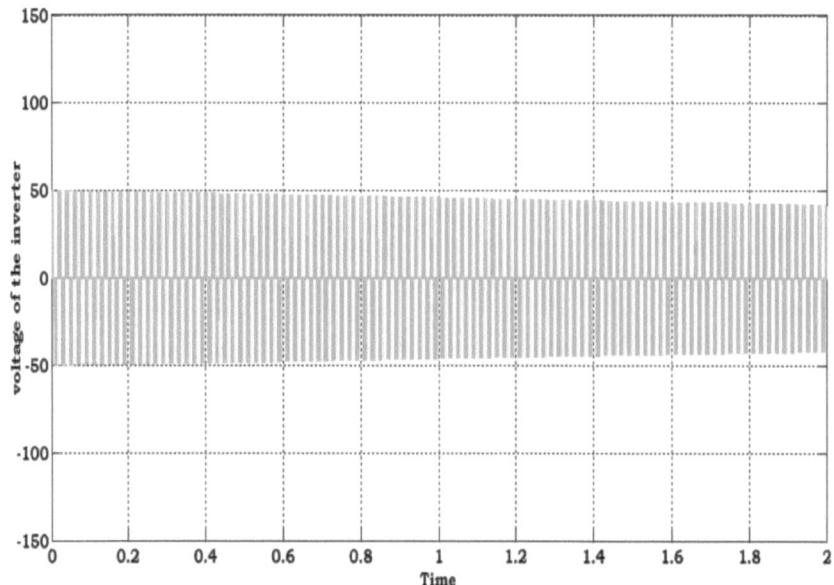

FIGURE 3.18 Output voltage of inverter.

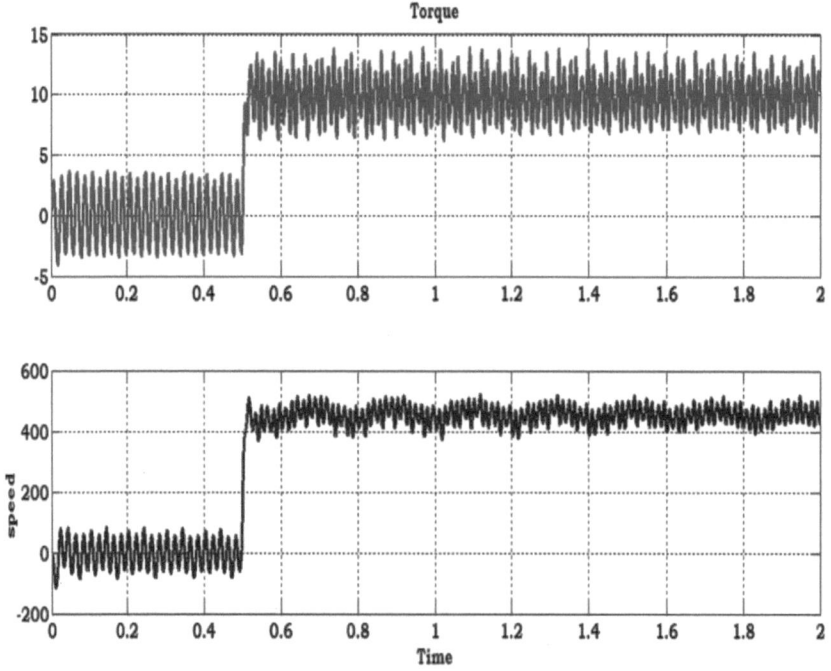

FIGURE 3.19 Speed, torque waveforms of BLDC motor.

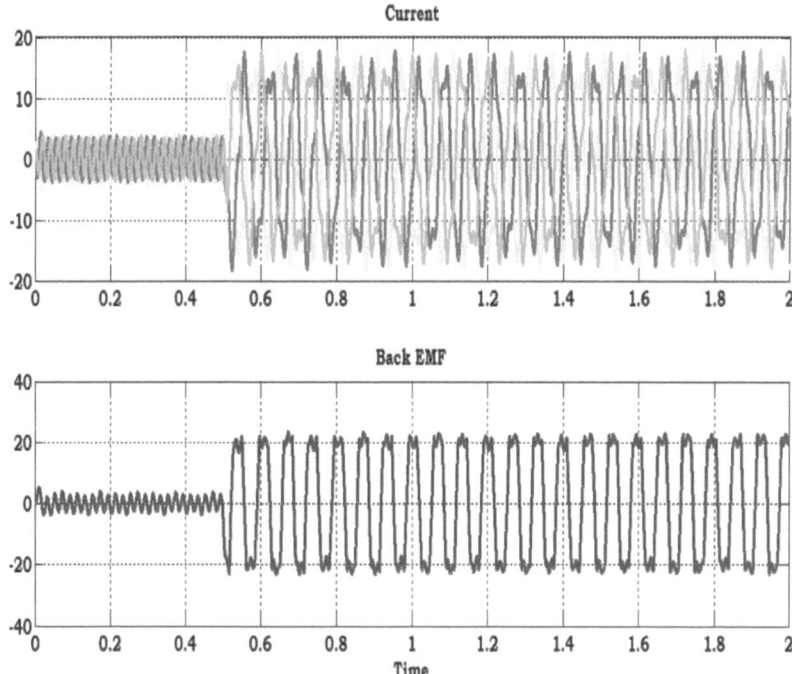

FIGURE 3.20 Current, EMF waveforms of BLDC motor.

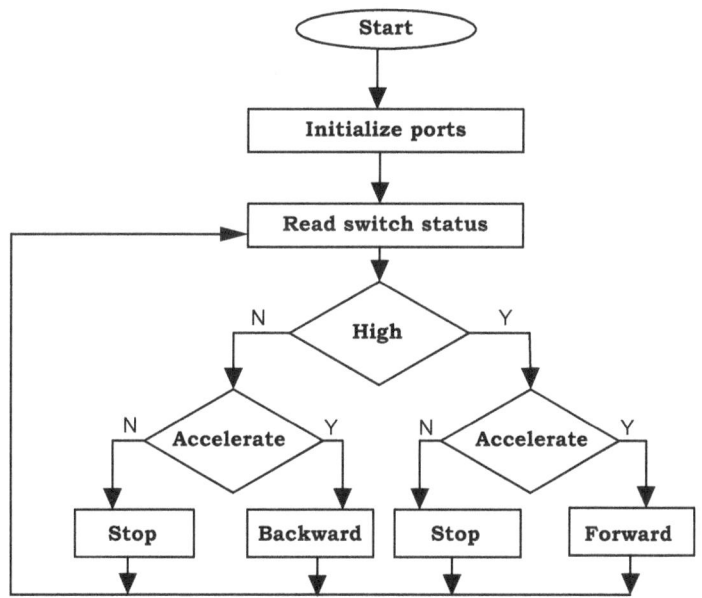

FIGURE 3.21 Flowchart of the vehicle control unit.

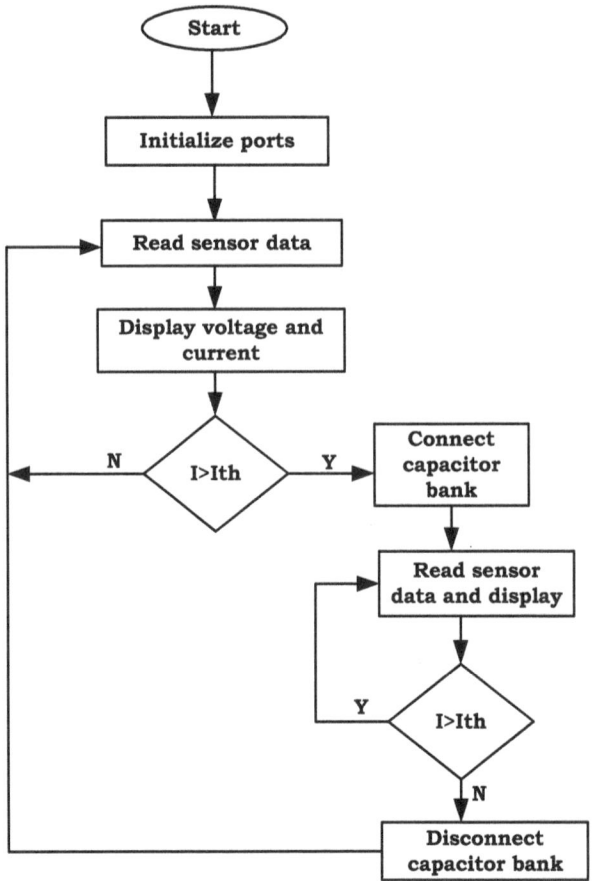

FIGURE 3.22   Flowchart of the vehicle capacitor bank control unit.

- All the ports will be initialized and subsequently switch status will be considered as the input and the voltage, current have to be displayed.
- Then, the threshold current will be compared with the actual current.
- If the threshold current is greater than the actual current, capacitor bank will be attached and then again the data will be read and displayed
- And if this threshold current is lesser than the actual current, again the data will be read.
- Then again verify if the threshold current is greater than the actual current and if this will not happen, disconnect the capacitor bank.

The flowchart for the vehicle block diagram is in shown Figure 3.23.

- The output from the solar panel will be fed to the charge controller and then to the battery for storage.
- The UC will also be charged with the output of the solar panel.

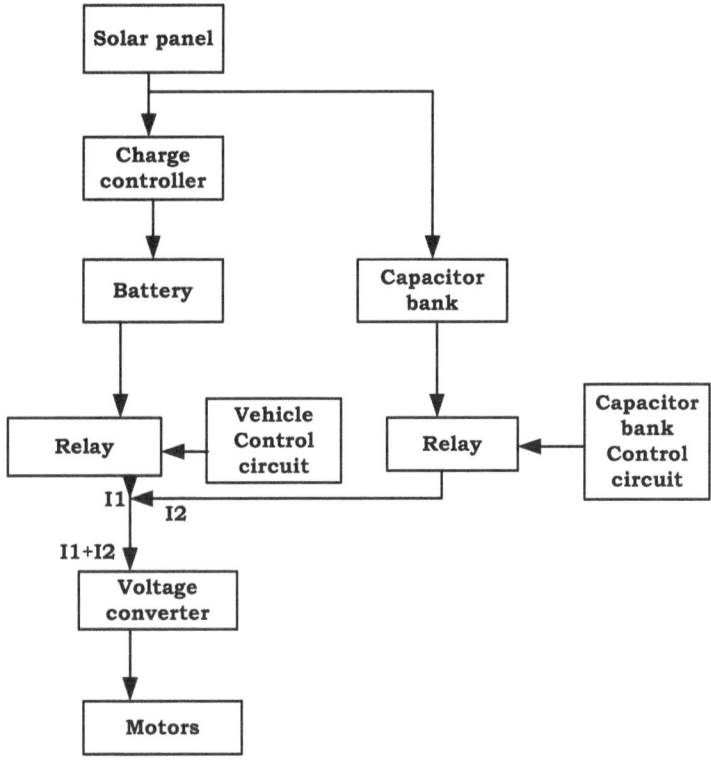

FIGURE 3.23   Flowchart of the vehicle operation.

- The battery current and the capacitor bank current will be fed to the voltage converter where the boosting or buck operation will be completed.
- Then, the boosted output will be fed to the inverter and then to the motor for driving the vehicle.

The proposed system has been implemented with the hardware model whose pictures are shown below which consists of energy storage system with UC, battery, DC–DC converter for converting the energy and BLDC motor for driving the vehicle (Figure 3.24).

The practical model contains the following components (Tables 3.2 and 3.3).

- Solar panel
- Charge controllers
- Relays
- PI controller
- DC–DC converter
- Battery
- UCs (Figures 3.25–3.27)

FIGURE 3.24 Proposed hybrid electric vehicle.

TABLE 3.2
Specifications of the Proposed Hybrid Vehicle

S. No	Parameter	Value
1	Seating capacity	4
2	Batteries (42AH/12V)	4
3	BLDC motors (240 W)	2
4	Solar panels (100 W/24V)	2
5	Ultra capacitors (1.5 f, 5.5V)	20
6	Charging time	5h
7	Full charge mileage	40km
8	Maximum speed	40 km/h
9	Chargeable	Dual source

TABLE 3.3
Representation of Data for Load Curve

Variable	Daytime	Evening
Time (h)	2.5 (8.00–10.30 A.M.)	2.5 (6.00–8.30 P.M.)
Distance (km)	25	18
Max. speed (kmph)	40	43
Average speed (kmph)	41	42
No. of stops	10	17
Stoppage time (min)	5	9

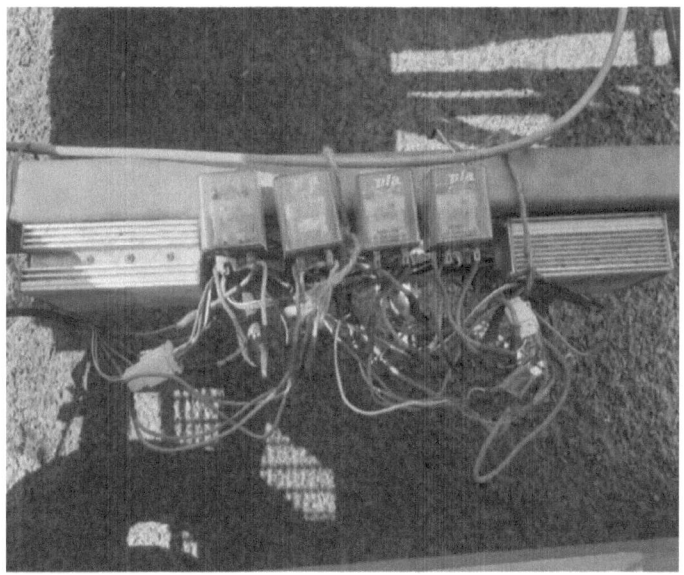

FIGURE 3.25 Representation of relays and solar charging.

FIGURE 3.26 Steering, pedal for acceleration and regeneration and arrangement of hub.

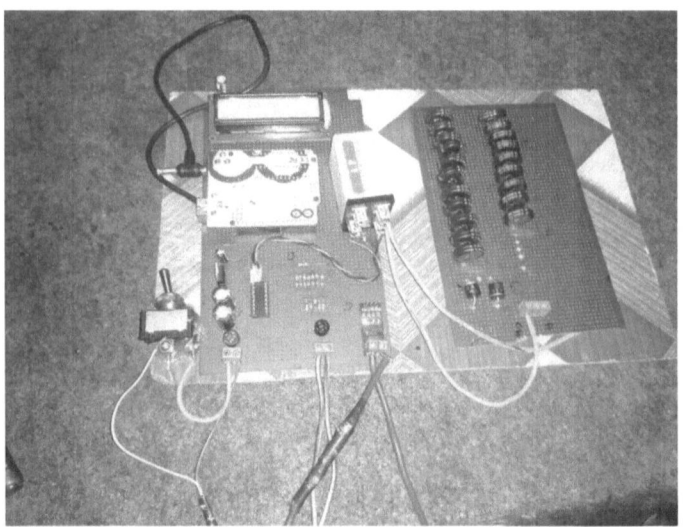

FIGURE 3.27   Ultra capacitor bank.

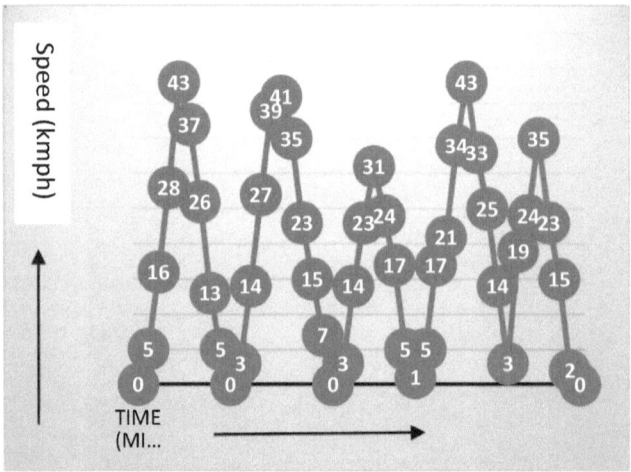

FIGURE 3.28   Speed vs time curve of hybrid vehicle without ultra capacitor.

The below graph shows the performance of the vehicle without UC which shows that there is a decrease in the speed up to zero speed due to variations in the solar irradiations (Figure 3.28).

The below graph shows the performance of the vehicle with UC which shows there is a decrease in the speed but nonzero speed due to variations in the solar irradiations (Figure 3.29).

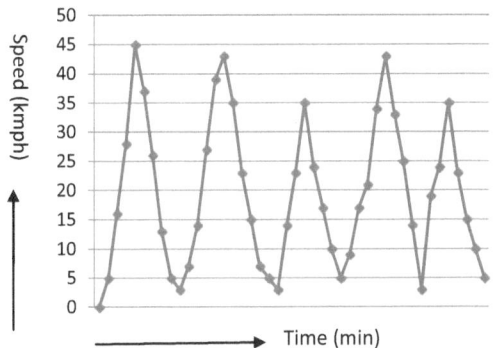

FIGURE 3.29 Speed vs time curve of hybrid vehicle with ultra capacitor.

3.7 CONCLUSIONS

The hybrid electric vehicle with UC, battery energy storage system for operating in the transient conditions has been presented in this chapter. The solar panel as an input has been provided for the hybrid vehicle and the two input full bridge buck boost converter has been designed with MOSFET switches as the DC–DC converter where the buck or boost operation can be obtained based on the input voltage. To control the output of the converter, a PI controller has been designed as the feedback controller. The entire design has been simulated using MATLAB and Simulink, and a hardware circuit consisting of a Arduino-based power circuit in an interfacing battery pack and super capacitors has been implemented.

REFERENCES

1. Burke, "Ultracapacitors: Why, how, and where is the technology," *Journal of Power Sources*, Vol. 91, pp. 37–50, 2000.
2. A.K. Shukla, S. Sampath and K. Vijaya mohanan, "Electrochemical super capacitors: Energy storage beyond batteries," *Current Science*, Vol. 79, p. 12, 25 December 2000.
3. C. Ashtiani, R. Wright and G. Hunt, "Ultra capacitors for automotive applications," *Journal of Power Sources*, Vol. 154, pp. 561–566, 2006.
4. Paul BowlesHuei PengHuei PengXiaobing Zhang, "Energy management in a parallel hybrid electric vehicle with a continuously variable transmission," Proceedings of the American Control Conference 1(6), pp. 55–59, October 2000.
5. T. Stuart, "Circuit with a Switch for Charging a Battery in a Battery Capacitor Circuit," Patent Application, The University of Toledo, 2004.
6. J. Miller, P. McClear and M. Cohen, "Ultra capacitors as Energy Buffers in a Multiple Zone Electrical Distribution System," Maxwell Technologies Inc. White Paper, www.maxwell.com.
7. D. Napoli, et al., "Design and testing of a fuel-cell powered propulsion system supported by a hybrid UC-battery storage," *SAE Journal*, pp. 151–160, 2004.
8. R. Dougal, S. Liu and R. White, "Power and life extension of battery-ultra capacitor hybrids," *IEEE Transactions on Components and Packaging Technologies*, Vol. 25, No. 1, pp. 120–131, March 2002.

9. F. Gagleardi, et al., "Experimental Results of on-board Battery-Ultra capacitor System for Electric Vehicle Applications," IEEE, 2002.

10. J. Miller and S. Butler, "Diesel engine starting using battery-capacitor combinations," *SAE Journal*, p. 10, 2001.

11. R. Schupbach, et al., "Design Methodology of a Combined Battery-Ultra capacitor Energy Storage Unit for Vehicle Power Management," *IEEE 2003 Power Electronics Specialists Conference (PESC)*, June 16th, Vol. 1, pp. 88–93, 2003.

12. A. Trippe, A. Burke and E. Blank, "Improved Electric Vehicle Performance with Pulsed Power Capacitors," *Proceedings of the ESD Environmental Vehicles Conference and Exposition*, Dearborn, MI, 1995.

13. J. Dixon and M. Ortuzar, "Ultra capacitors + DC-DC converters in regenerative braking system," *IEEE AESS Systems Magazine*, August 2002.

14. J. Mierlo, P. Bossche and G. Maggetto, "Models of energy sources for EV and HEV: Fuel cells, batteries, ultra capacitors, flywheels, and engine generators," *Journal of Power Sources*, Vol. 128, pp. 76–89, 2003.

15. R. Dougal, L. Gao and S. Liu, "Ultra capacitor model with automatic order selection and capacity scaling for dynamic system simulation," *Journal of Power Sources*, Vol. 126, pp. 250–257, 2003.

16. A. Chu and P. Braatz, "Comparison of commercial super capacitors and high-power lithium-ion batteries for power-assist applications in hybrid electric vehicles I. Initial Characterization," *Journal of Power Sources*, Vol. 112, pp. 236–246, 2002.

17. L.T. Lam, N.P. Haigh, C.G. Phyland and A.J. Urban, "Failure mode of valve-regulated lead-acid batteries under high-rate partial-state-of-charge operation," *Journal of Power Sources*, Vol. 133, pp. 126–134, 2004.

18. P. Ruetschi, "Aging mechanisms and service life of lead-acid batteries," *Journal of Power Sources*, Vol. 127, pp. 33–44, 2004.

19. J.H. Yan, W.S. Li and Q.Y. Zhan, "Failure mechanism of valve-regulated lead-acid batteries under high-power cycling," *Journal of Power Sources*, Vol. 133, pp. 135–140, 2004.

20. R. Wagner, "Failure modes of valve-regulated lead/acid batteries in different applications," *Journal of Power Sources*, Vol. 53, pp. 153–162, 1995.

21. A. Cooper, "Development of a lead-acid battery for a hybrid electric vehicle," *Journal of Power Sources*, Vol. 133, pp. 116–125, 2004.

22. G.J. May, "Valve-regulated lead-acid batteries for stop-and-go applications," *Journal of Power Sources*, Vol. 133, pp. 110–115, 2004.

23. P.T. Moseley, "High rate partial-state-of-charge operation of VRLA batteries," *Journal of Power Sources*, Vol. 127, pp. 27–32, 2004.

24. A. Hande, "AC Battery Heating for Cold Climates," Ph.D. Dissertation, The University of Toledo, 2002.

4 Recreation and Control of Multi Rotor Wind Power Generating Scheme for Pumping Applications

M. Divya
Vignan's Lara Institute of Technology & Science

Geetha Reddy Evuri and K. Srinivasa Reddy
Teegala Krishna Reddy College of Engineering and Technology (TKRCET)

B. Nagi Reddy
Vignana Bharathi Institute of Technology (VBIT)

CONTENTS

4.1 INTRODUCTION

Wind is spotless, sustainable, boundless vitality hotspot for electric force age. The removed vitality from a wind turbine reliant upon gust speed in this mode is financially progressively gainful to introduce the breeze turbines [1–5] at districts where

DOI: 10.1201/9781003143802-4

the breeze power thickness is high. The removed vitality from a breeze turbine additionally relies upon the effectiveness of the breeze turbine. Wind turbines are made in a wide range of erect and smooth core. The littlest turbines are employed for relevance, for instance, battery charging for helper power for pontoons or bands or to control traffic notice signs. Larger turbines can be operated for making commitments to a residential force supply while selling unused force rear to service the source through the electrical matrix.

Multi rotor turbines that have two rotors, one at the front and one at the back of the breeze turbines, offers high productivity. A solitary rotor wind turbine hypothetically can extricate most extreme 59.3% of the vitality accessible in the breeze and this breaking point is known as far as possible. Utilizing numerous actuator-plate hypothesis, one can without much of a stretch show that a twofold rotor wind turbine hypothetically can separate most extreme 64% of the breeze. In this behavior the majority extreme vitality that a turbine can hypothetically extricate from the breeze is 8% higher than that of a solitary rotor turbine. In the double rotor turbine, the rotors can pivot either in co-bearing (rotors are turning in a similar way), or use oil in counter-heading (rotors are pivoting inverse course).

Counter turning double rotor wind turbines, by and large, can have diverse mechanical designs. For instance one methodology is to consolidate the counter revolution of the two rotors utilizing a gear framework and acquire a solitary rotational yield shaft which is associated with the pole of a generator. Right now, generator can be utilized [10–16] in the breeze turbine with no change since it is one mechanical turning shaft yield from the apparatus framework. Another methodology is to utilize an extraordinary generator approach, in which both the main player rotor and twisting stator of the generator can turn freely, thusly one rotor of the breeze turbine is associated with the central player rotor of the generator and the second rotor of the breeze turbine is associated with the twisting stator of the breeze turbine.

As a result as to boost up the proficiency of breeze turbines while limiting the load, wind turbines are worked leveled out. Since, in nature the breeze has constantly changing qualities with consistently shifting breeze speeds, the streamlined torque and intensity of the breeze turbine cutting edges. In a twofold turbine, the breeze first disregards front rotor edges and loses some part of its vitality. At that point the downstream of the front rotor, the remaining wind with lower energy ignores the back rotor sharp edges and loses some segment of its energy further.

4.2 BLOCK DIAGRAM AND MATHEMATICAL MODELING OF MULTI ROTOR WIND TURBINE

The block diagram of multi rotor wind turbine power generating system is shown in Figure 4.1.

4.2.1 WIND TURBINE

A wind turbine, or on the other hand alluded to as a breeze vitality converter, is an appliance that alters over breeze dynamic vitality into electrical vitality. Wind turbines [11–14] are produced in a wide scope of vertical and level pivot. The smallest

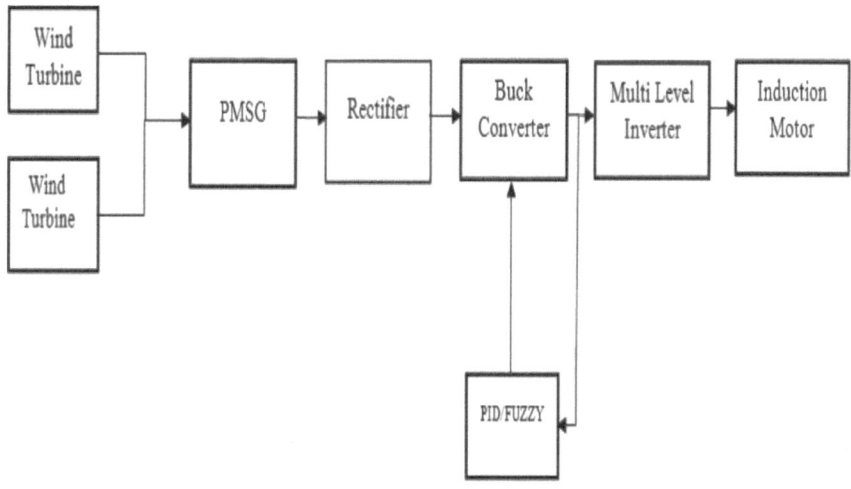

FIGURE 4.1 Block drawing of the multi rotor turbine system.

turbines are employed for functions, like charging of battery for pontoons or trains or to control traffic notice signs. Larger turbines can be employed for making commitments to a household power supply while selling unused force back to utility contributor through the electrical matrix. These turbines convert the wind energy into mechanical power with the rotation of blades.

$$\text{Output power } P_T = T_m \times \omega \tag{4.1}$$

$$\text{Kinetic energy } P_T = \frac{1}{2}\rho \times V^3 \times A \tag{4.2}$$

The energy retrieved from the stream can be represented as Table 4.1.

TABLE 4.1
Difference of SRWT and DRWT

S No	Specification	Single Rotor	Dual Rotor
1	Gearbox arrangement	It may be planetary or parallel shaft gearbox.	This wind turbine gearbox is the combination of both planetary and parallel shaft
2	Power and torque	In sole rotor wind turbine, the wind energy we received is equal.	Twofold turbine has an advantage of extra energy capture. The energy which is unused by the front rotor can be utilized by the rear rotor.
3	Torque	The torque produced will be less in the single rotor wind turbine.	We get double torque from front and rear rotors.

$$C_p = \frac{1}{2} \times \rho A V \left(V_o + V_1 \right) \left(V_o - V_1 \right) \tag{4.3}$$

$$C_p = \frac{\rho}{0.5 \times V_o^3 \times A} \tag{4.4}$$

$$\lambda = \frac{r \times \omega}{V_o} \tag{4.5}$$

4.2.2 Permanent Magnet-Based Synchronous Generator

The name synchronous alludes here that rotor and attractive field pivot with a similar speed, in light of the truth that field is produced through a pole mounted lasting magnet instrument and current is instigated into the stationary armature.

$$C_p = \frac{Tm \times \omega}{\left(1/2\rho\right) \times V^3 \times A} \tag{4.6}$$

The wind turbine mechanical characteristics

$$T_m - T_g = J \frac{d\omega}{dt} \tag{4.7}$$

4.2.3 Diode Bridge Rectifier

A rectifier is an electrical apparatus that alter more alternating flow (AC), to direct flow (DC), with the help of diodes. Genuinely, rectifiers take various structures, including vacuum tube diodes, wet synthetic cells, mercury-bend valves, piles of copper and selenium oxide plates, other silicon-based semiconductor switches.

4.2.4 DC–DC Converter

A buck converter has been used for altering the higher voltage into lower rated voltage with the use of switch, diode, inductor, and capacitor.

$$\text{Maximum duty cycle } D = \frac{V_{\text{out}}}{V_{\text{in}}(\text{max}) \times \eta} \tag{4.8}$$

$$\text{Inductor } L = \frac{V_{\text{out}}\left(V_{\text{in}} - V_{\text{out}}\right)}{\Delta Il \times fs \times V_{\text{in}}} \tag{4.9}$$

$$C_{\text{OUT}}(\text{min}) = \frac{\Delta Il}{8 \times fs \times \Delta V_{\text{out}}} \tag{4.10}$$

4.2.5 PID CONTROLLER

A proportional fundamental subordinate regulator is a nonexclusive control feedback criticism instrument broadly utilized in mechanical control frameworks. A PID is the most regularly utilized regulator. A PID control ascertains blunder esteem as the contrast between a deliberate cycle variable and an ideal set point. The regulator endeavors to limit the blunder by altering the cycle control inputs. The PID regulator count includes three separate consistent boundaries and once in a while called three-term control: the relative, the vital and subordinate qualities, meant P, I, D. Heuristically, these qualities can be deciphered as time: P relies upon the current blunder, I on the gathering of past mistakes, and D is a forecast future blunders, in light of the current mistake, in view of the current pace of progress.

4.2.6 FUZZY CONTROLLER

A fuzzy control framework is a control framework dependent on the fuzzy rationale, which is a numerical framework that examines simple information esteems as far as intelligent factors that take on ceaseless qualities somewhere in the range of 0 and 1, as opposed to old style or advanced rationale, which works on discrete estimations of either 1 or 0 (valid or bogus, separately) (Figure 4.2).

TABLE 4.2
Rule Base for Fuzzy Controller

Error/Change in Error	NH	NL	ZE	PL	PH
PH	ZE	PL	PH	PH	PH
PL	NL	ZE	PL	PH	PH
ZE	NH	NL	ZE	PL	PH
NL	NH	NH	NL	ZE	PL
NH	NH	NH	NH	NL	ZE

Where
    PH = Positive High
    PL = Positive Low
    NH = Negative High
    NL = Negative Low
    ZE = Zero Equal

4.3 SIMULINK MODEL OF DOUBLE ROTOR WIND TURBINE

In dual rotor-supported wind turbine, two wind turbines are worn. To advance an output voltage value of all the energy sources, the multiple input buck is used where the output voltage is step down to 600 V. To avoid the oscillations in the output voltage of the buck controllers are used. Controllers like PID and fuzzy are used (Figure 4.2).

Figure 4.3 shows the output speed of the single rotor wind turbine.

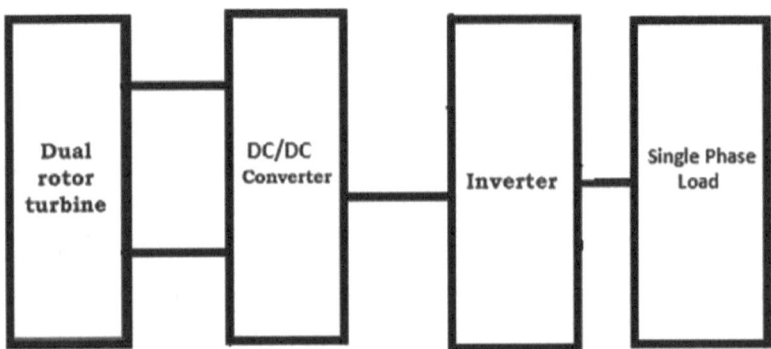

FIGURE 4.2 Simulink model of the dual twofold rotor wind turbine.

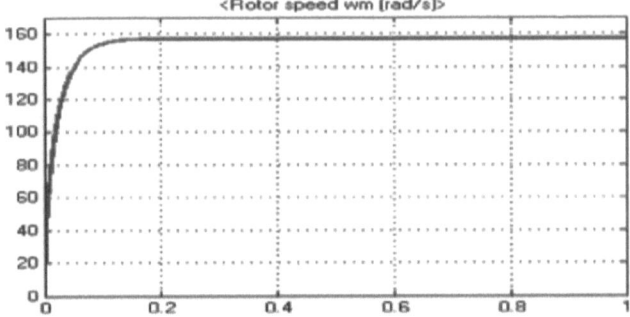

FIGURE 4.3 Speed output of the single rotor wind turbine.

Figure 4.4 shows the output torque of the single rotor wind turbine.
Figure 4.5 shows the voltage output waveform of the single rotor wind turbine.
Figure 4.6 shows the current output waveform of the single rotor wind turbine.
Figure 4.7 illustrates the speed output waveform of the twofold rotor turbine.

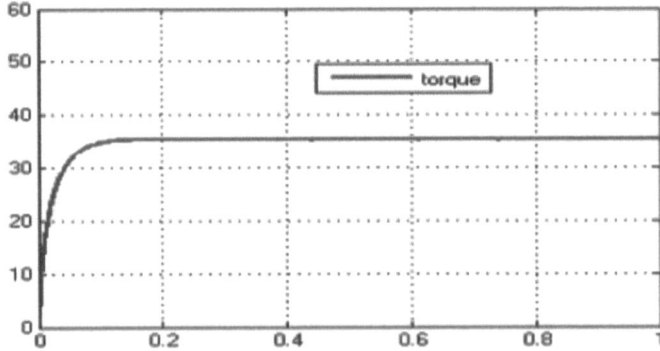

FIGURE 4.4 Torque of the single rotor wind turbine.

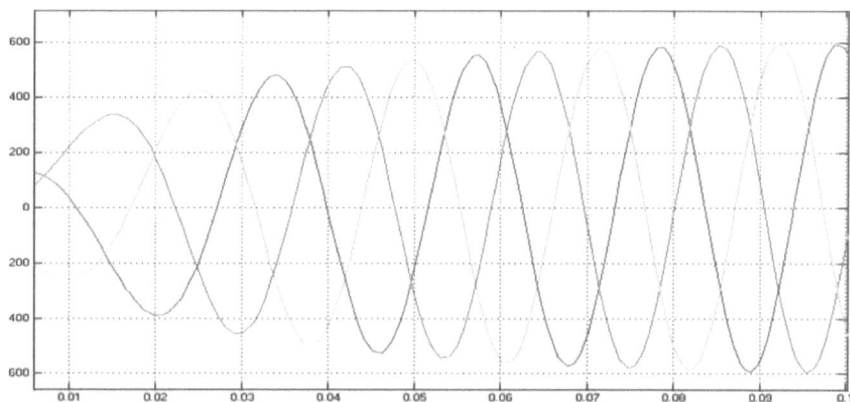

FIGURE 4.5 Voltage output waveform of the single rotor wind turbine.

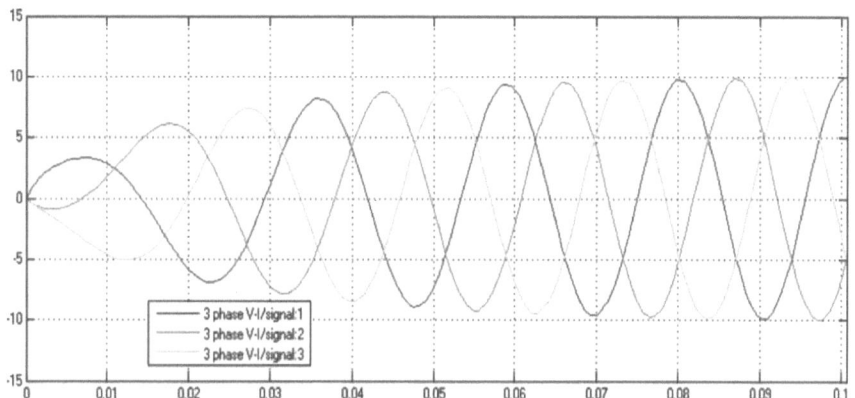

FIGURE 4.6 Output current for the sole rotor turbine.

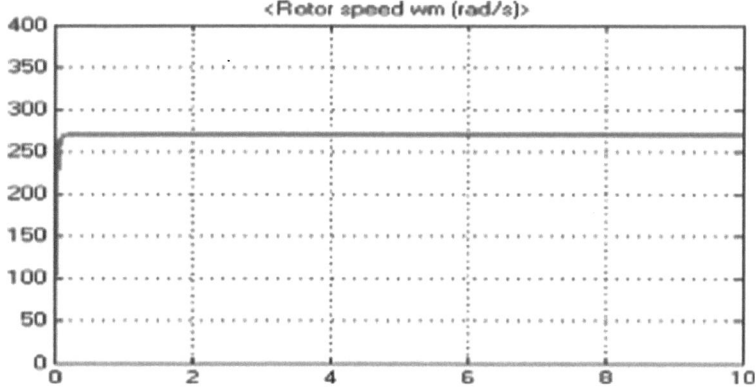

FIGURE 4.7 Speed for the twofold rotor turbine.

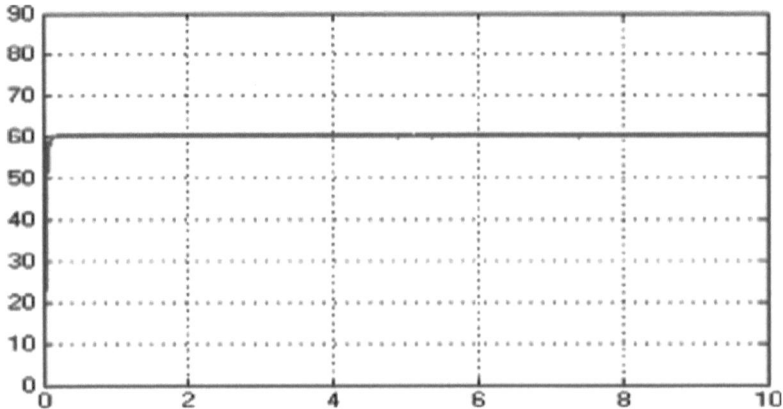

FIGURE 4.8 Torque for the twofold rotor turbine.

Figure 4.8 shows the output of the torque waveform of the twofold rotor turbine.
Figure 4.9 shows the voltage output waveform of the twofold rotor turbine.
Figure 4.10 shows the current output waveform of the twofold rotor turbine.
The underneath Figure 4.11 illustrates the potential waveform of the rectifier.
Figure 4.12 shows the potential of the buck converter.
Figure 4.13 shows the current waveform of the buck converter.
Figure 4.14 shows the simulation of the output current for buck with PID controller.
The beneath waveform shows simulation of the output voltage for buck with PID (Figure 4.15).
Figure 4.16 shows the potential of the fuzzy logic controller with twofold turbine.

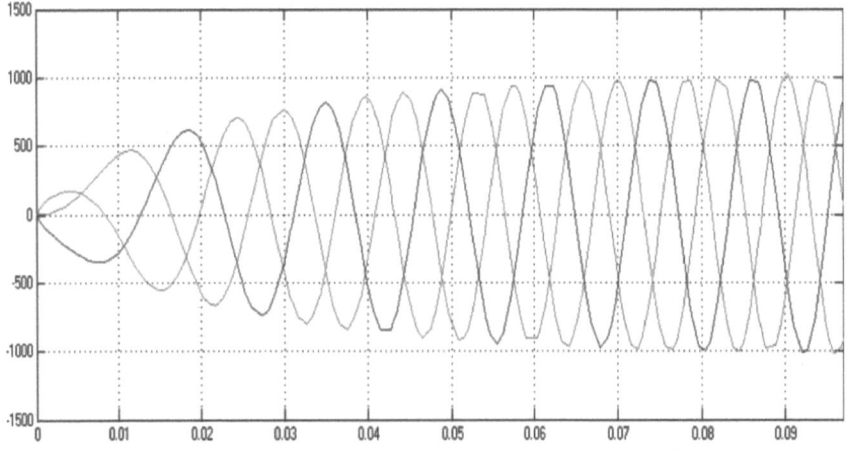

FIGURE 4.9 Output voltage for the twofold rotor turbine.

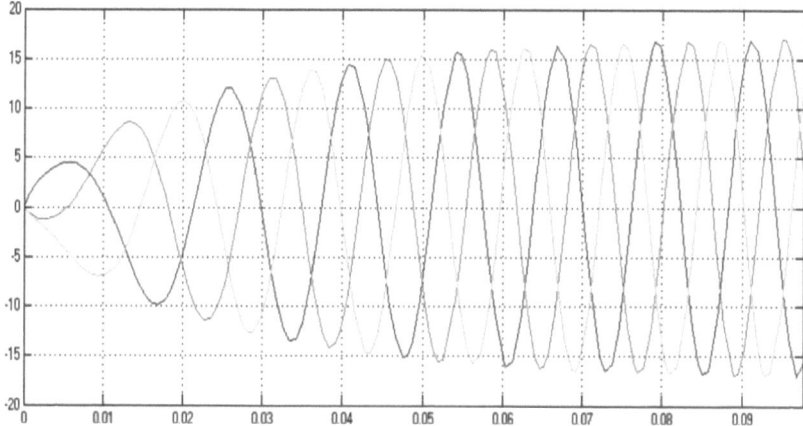

FIGURE 4.10 Output current for the twofold rotor turbine.

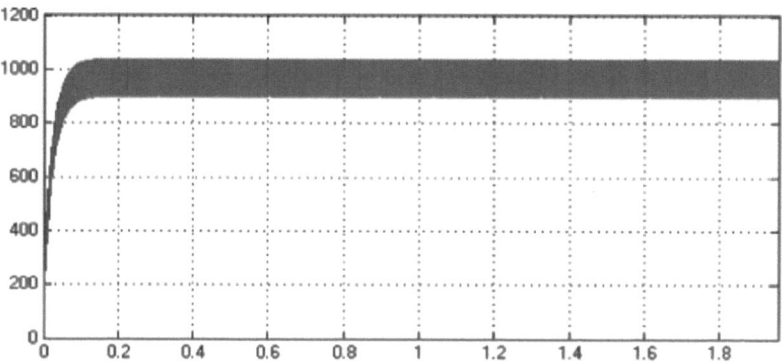

FIGURE 4.11 Output potential for the rectifier.

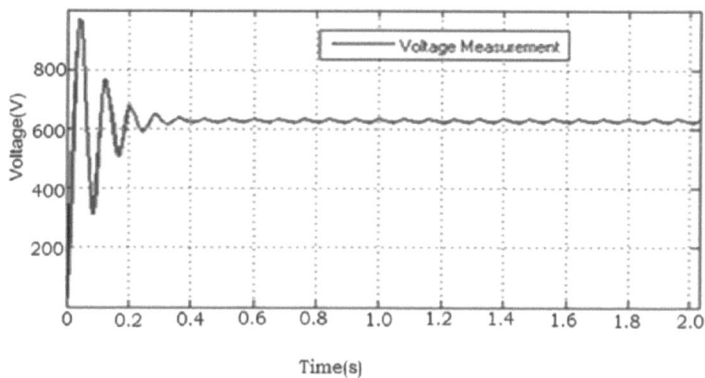

FIGURE 4.12 Output potential for the buck converter.

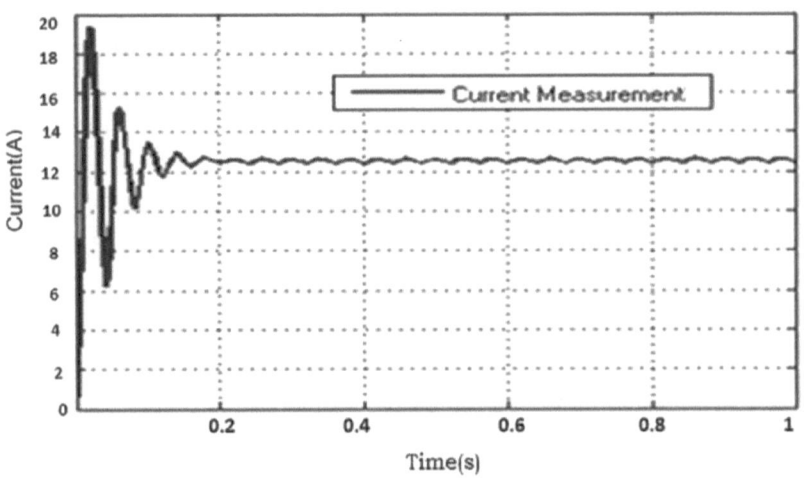

FIGURE 4.13   Current for the buck converter.

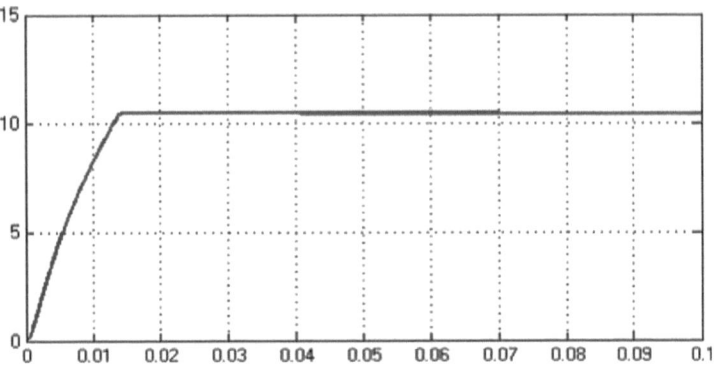

FIGURE 4.14   Current output for buck with the PID controller.

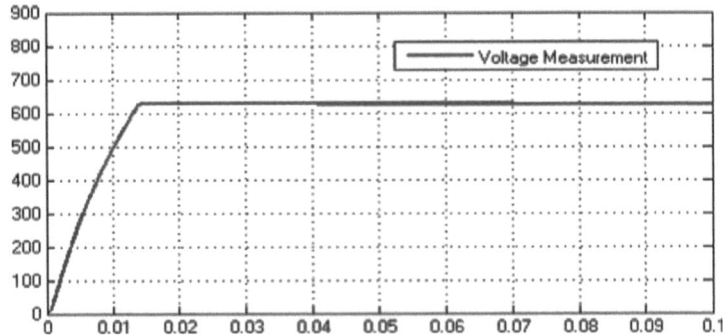

FIGURE 4.15   Potential for buck with the PID controller.

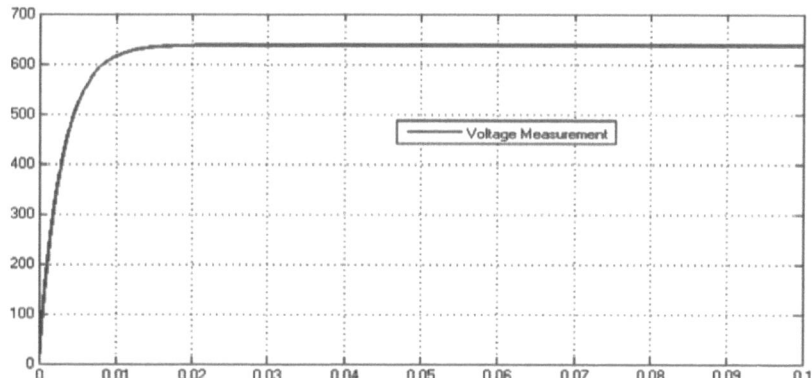

FIGURE 4.16 Output voltage of the fuzzy logic controller with twofold turbine.

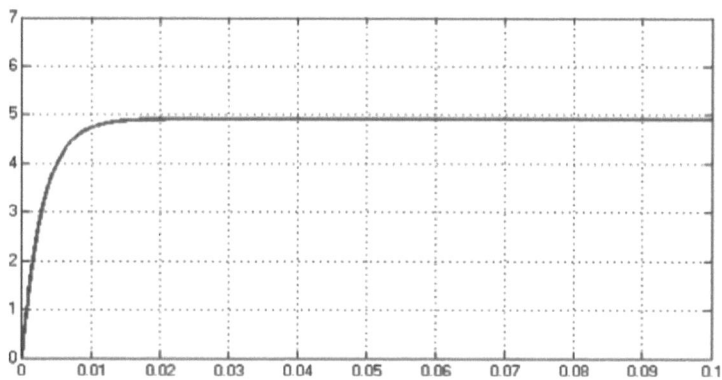

FIGURE 4.17 Output current of the fuzzy logic controller with twofold turbine.

TABLE 4.3
Parameters of SRWT and DRWT

S. No.	Parameter	Sole Rotor Turbine	Dual Fold Turbine
1	Speed	1,500	2,511
2	Torque	31.8	60
3	Voltage	587	972
4	Current	9.93	17.28

Figure 4.17 shows the output current of the fuzzy logic controller with twofold turbine (Tables 4.3 and 4.4).

Figure 4.18 shows the phase voltage of the inverter.

Figure 4.19 shows the line potential of the five level inverter.

Figure 4.20 gives the speed of the motor.

Figure 4.21 gives the torque of the motor (Tables 4.5 and 4.6).

TABLE 4.4

Comparison of Performance of Controllers

Parameter	Without Controller	With Controller	With Fuzzy
Rise time (s)	0.02	0.01	0.005
Peak time (s)	0.04	0.015	0.01
Settling time (s)	0.02	0.02	0.014

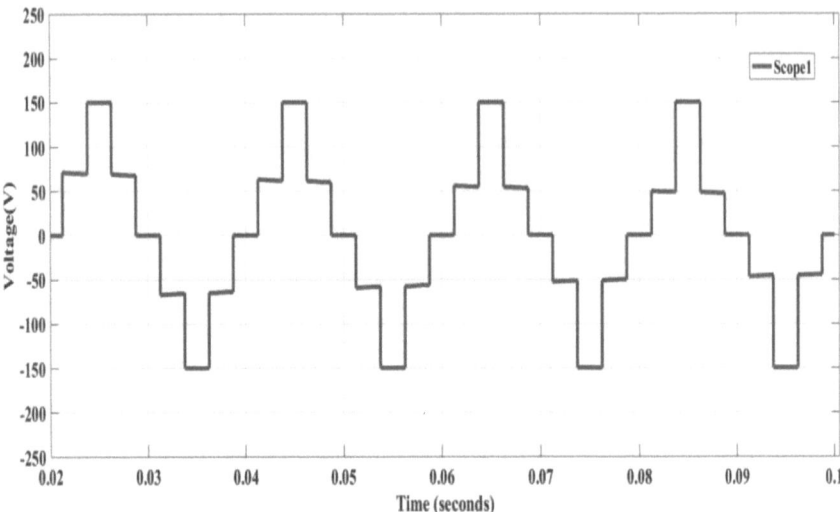

FIGURE 4.18 Output phase potential of the multilevel inverter.

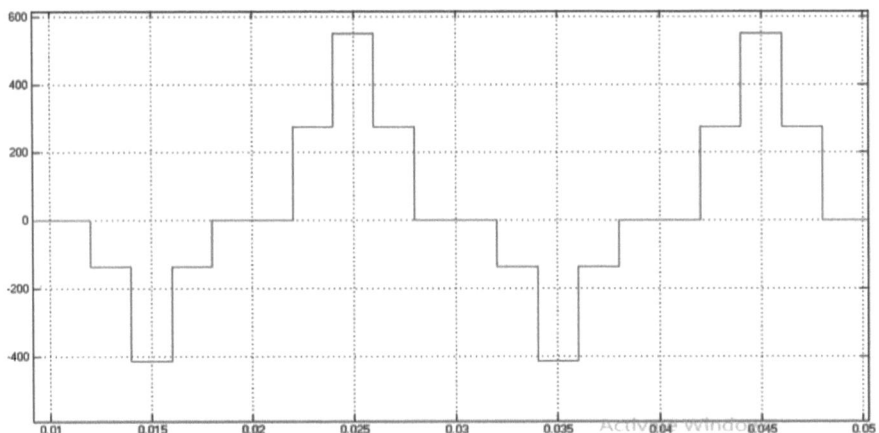

FIGURE 4.19 Line potential of the multilevel inverter.

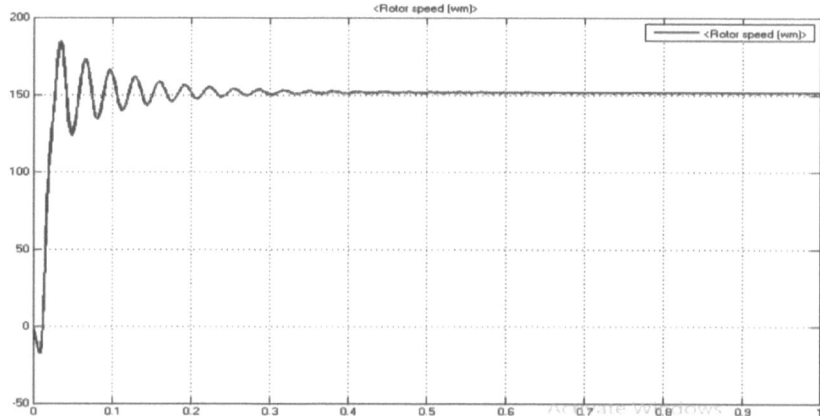

FIGURE 4.20   Speed output of the five multilevel inverter.

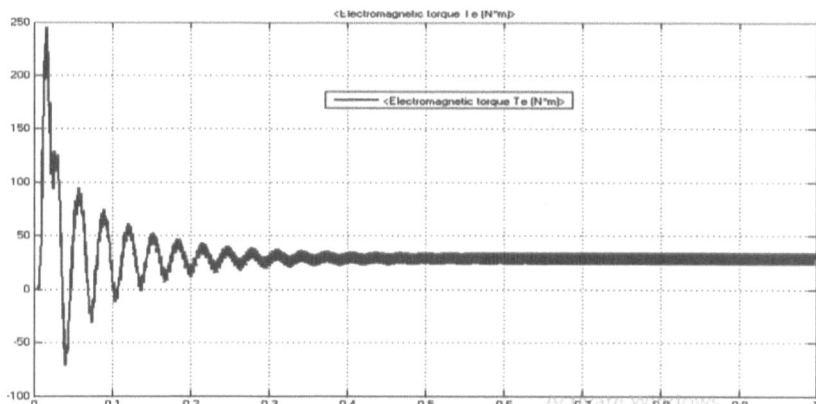

FIGURE 4.21   Torque output for the five level inverter with induction motor.

TABLE 4.5
Shows the Specifications of Dual Rotor Wind Turbine

S. No	Parameter	Symbol	Value	Units
1	Wind power	P	10.9	kW
2	Current	I	17.83	A
3	Voltage	V	972	V
4	Stator resistance	R_s	300	Ω
5	Rotor resistance	R_r	225	Ω
6	Stator inductor	L_s	120	μH
7	Rotor inductance	R_s	50	μH
8	Speed	N	2,511	RPM
9	Torque	T	60	N-M

TABLE 4.6
Specifications of Buck Converter

S. No	Parameter	Symbol	Value	Units
1	Resistance	R	0.001	Ω
2	Inductance	L	80×10^{-6}	H
3	Capacitor	C	90×10^{-6}	F

4.4 CONCLUSIONS

The two rotor-based airstream turbine generator scheme, consisting of two rotors, has been represented in this chapter. Also, PMSG for producing torque and power has been intended by means of the second rotor and there has been an increase in the power output of the dual rotor wind turbine. Simulation results provide that the sole rotor scheme is further receptive to torque oscillations. Speed fluctuation in the sole system is larger than that in the twofold system under transient conditions. A buck converter has been designed with two controllers named as PID controller and fuzzy controller and an assessment has been prepared between the controllers with respect to settling time, rise time and peak time where the converter with the fuzzy controller shows the better performance.

4.5 FUTURE WORK

The dual rotor wind turbine can be executed with different types of motors and their performance can be improved by the induction motor (or) BLDC motor. It can also be extended with different types of multilevel inverters to reduce the output harmonics of the system.

The scope of wind power improves the utilization efficiency of resources, helps wind power generation completely replace traditional power generation, better reduces pollution emissions, saves fossil energy, has broad application prospect, and provides a new direction for wind power construction in the future.

REFERENCES

1. C. Rajendran and G. Madhu, "A CFD based multibody dynamics approach in horizontal axis wind turbine," *International Journal of Dynamics of Fluids*, vol. 6, no. 2, pp. 219–230, 2010.
2. I. Grant, M. Mo, X. Pan et al., "Experimental and numerical study of the vortex filaments in the wake of an operational, horizontal-axis, wind turbine," *Journal of Wind Engineering and Industrial Aerodynamics*, vol. 85, no. 2, pp. 177–189, 2000.
3. J. Sumner, C.S. Watters, and C. Masson, "CFD in wind energy: The virtual, multiscale wind tunnel," *Energies*, vol. 3, no. 5, pp. 989–1013, 2010.
4. L.J. Vermeera, J.N. Sorensenb, and A. Crespo, "Wind turbine wake aerodynamics," *Progress in Aerospace Sciences*, vol. 39, no. 6–7, pp. 467–510, 2003.

5. P.R. Ebert and D.H. Wood, "The near wake of a model horizontal-axis wind turbine-I. Experimental arrangements and initial results," *Renewable Energy*, vol. 12, no. 3, pp. 225–243, 1997.
6. B.G. Newman, "Actuator-disc theory for vertical-axis wind turbines," *Journal of Wind Engineering and Industrial Aerodynamics*, vol. 15, no. 1–3, pp. 347–355, 1983.
7. B.G. Newman, "Multiple actuator-disc theory for wind turbines," *Journal of Wind Engineering and Industrial Aerodynamics*, vol. 24, no. 3, pp. 215–225, 1986.
8. W.Z. Shen, V.A.K. Zakkam, J.N. Sørensen, and K. Appa, "Analysis of counter-rotating wind turbines," *Journal of Physics: Conference Series*, vol. 75, no. 1, pp. 1–9, 2007.
9. S.N. Jung, T.S. No, and K.W. Ryu, "Aerodynamic performance prediction of a 30 kW counter-rotating wind turbine system," *Renewable Energy*, vol. 30, no. 5, pp. 631–644, 2005.
10. J.D. Booker, P.H. Mellor, R. Wrobel, and D. Drury, "A compact, high efficiency contrarotating generator suitable for wind turbines in the urban environment," *Renewable Energy*, vol. 35, no. 9, pp. 2027–2033, 2010.
11. W. Cho, K. Lee, I. Choy, and J. Back, "Development and experimental verification of counter-rotating dual rotor/dual generator wind turbine: Generating yawing and furling," *Renewable Energy*, vol. 114, pp. 644–654, 2017.
12. A. Yahdou, B. Hemici, and Z. Boujema, "Sliding mode control of dual rotor wind turbine system," *The Mediterranean Journal of Measurement and Control*, vol. 11, no. 2, pp. 412–419, April 2015.
13. E.M. Farahani, N. Hosseinzadeh, and M.E. Mehran, "Comparison of dynamic responses of dual and single rotor wind turbines under transient conditions," *2010 IEEE International Conference on Sustainable Energy Technologies (ICSET)*, Kandy, 2010, pp. 1–8.
14. R.W.Y. Habash, V. Groza, and P. Guillemette, "Performance optimization of a dual-rotor wind turbine system," *2010 IEEE Electrical Power & Energy Conference*, Halifax, NS, 2010, pp. 1–6.
15. T.S. No, J.-E. Kim, J.H. Moon, and S.J. Kim, "Modeling control and simulation of dual rotor wind turbine generator system," *Renewable Energy*, vol. 34, pp. 2124–2132, 2009.
16. K. Nounou and K. Marouani, "Control of a dual star induction generator driven wind turbine," *2016 19th International Symposium on Electrical Apparatus and Technologies (SIELA)*, Bourgas, 2016, pp. 1–4.
17. A. Tapia, G. Tapia, J.X. Ostolaza, and J.R. Sáenz, "Modeling and control of a wind turbine driven doubly fed induction generator," *IEEE Transactions on Energy Conversion*, vol. 18, no. 2, pp. 194–204, June 2003.

5 Analytical Study of DC-to-DC Power Converter

Srikanth Goud B
Anurag College of Engineering

Ankit Kumar
Hanyang University

Shyam Akashe
ITM University Gwalior

Seelam VSV Prabhu Deva Kumar
JoongAng Control Co., Ltd.

Ch Rami Reddy
Nehru Institute of Engineering and Technology

CONTENTS

DOI: 10.1201/9781003143802-5

5.1 INTRODUCTION

Modern science is always looking for upgrading techniques used and this can be done by comparing the devices and machinery that were used a few decades ago. Similarly, ever since electronic devices have come into picture, one of the basic issues for the developer is to reduce the time and numbers of stages for power conversion. Power converter plays an important role when a system is to be made.

As per its definition, it is a device used for converting electrical energy from

a. Alternating Current (AC)–Direct Current (DC)
b. Alternating Current (AC)–Alternating Current (AC)
c. Direct Current (DC)–Direct Current (DC)
d. Direct Current (DC)–Alternating Current (AC)

Nowadays, a modern power converter undergoes many stages of conversion. Here, this chapter is restricted to DC/DC converter in which DC voltage at the input side is converted to another DC voltage at the output side, which may be low or high to that of given DC input source voltage.

Furthermore, the DC/DC power converter is distinguishing into isolated power converter and non-isolated power converter. The isolated power converter has transformer in between the input and output as a barrier. Irrespective of the isolated power converter, the non-isolated power converter has no such any kind of barrier means the DC path is between its input and output [1,2].

These isolated power converter and non-isolated power converter are again sub-divided into different types. Figure 5.1 gives the classification of the DC/DC converter.

From Figure 5.1, the isolated power converter is classified into a resonant, bridge, forward and flyback. Similarly, a non-isolated power converter is categorized into buck converter, single-ended primary inductance converter (SEPIC converter), boost converter, zeta converter, and buck-boost converter.

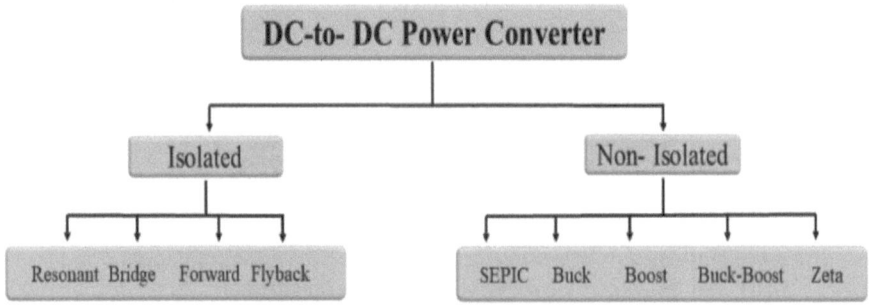

FIGURE 5.1 Classification of DC–DC power converter.

The resonant power converter has an L–C network in which sinusoidally the parameter voltage and current waveforms vary at each switching period intervals. Bridge converter is categorized into full- and half-bridge converter, respectively, where active switching components are either four or two in numbers and are configured in a bridge across converter [3,4]. A forward converter circuit is the reverse of a flyback converter where the energy from the primary to the secondary winding gets transferred while the switch is ON [5]. The zeta converter has a positive voltage at its output side drawn from the voltage at its input side in which output voltage differs [6]. Buck power converter steps down output voltage and boost power converter steps up voltage at its output.

This chapter is solely focused on the flyback and buck-boost power converter where it is divided into different sections where comparison, findings of both the converter is made and at last conclusion is drawn.

5.2 FLYBACK POWER CONVERTER

The flyback converter falls in the category of DC–DC power converter where the barrier is present between the input and output in which not the barrier but the transformer provides isolation. The flyback converter works in two different modes,

- Continuous
- Discontinuous

The function is the same as the buck-boost power converter except isolation is given in the flyback converter input and output with the help of the ferrite core high-frequency transformer and has multiple outputs [7]. The winding of the transformer is carried out in the opposite direction such that if the current given in primary winding will flow then current in secondary winding has zero current and vice-versa [7].

5.2.1 FLYBACK POWER CONVERTER IN THE DISCONTINUOUS MODE

The discontinuous mode of the flyback converter has three modes. These are explained below.

Mode I: Switch T is ON, and current $IS = 0$

In this mode, the switch T is in ON state, then the current going to flow in the Loop 1, the primary winding current IP will rise gradually and the voltage VT across switch T is 0. At the same, the secondary winding current IS will be 0. Since current IP increases gradually, this results in a gradual increase in flux too. The output voltage will be negative of the Edc.

Mode II: Switch T is OFF, and current $IP = 0$

In mode II, when the switch is in the OFF state, then the polarity in the primary winding will be reversed and the voltage across the switch T is twice the input voltage Edc. Due to the inductance of the transformer, the spike will arise which can be seen in Figure 5.2. The current IP in the primary winding will become 0 and the current IS will flow through the

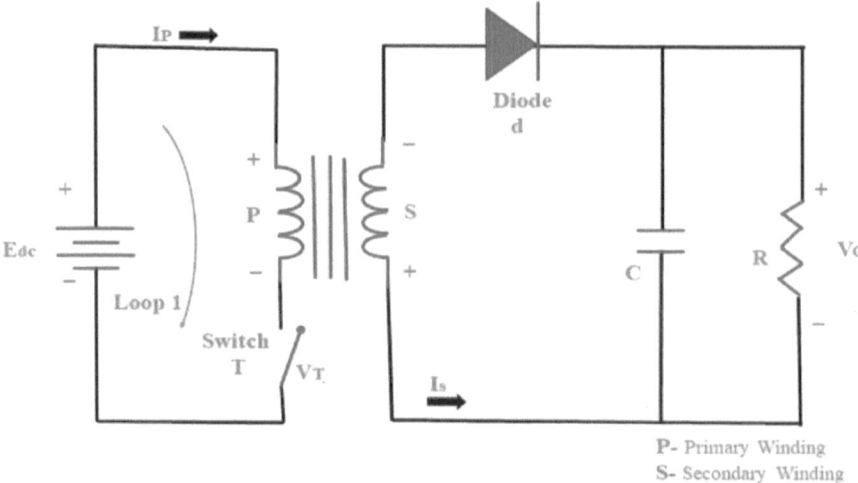

P- Primary Winding
S- Secondary Winding

FIGURE 5.2   The circuit diagram of the flyback power converter in the discontinuous mode.

secondary winding across capacitor C and load resistor R which will gradually decrease. At the same time, the flux of secondary winding will also
decrease gradually. When the current IS will become 0 and still, the switch
T is in the OFF state, then the input voltage Edc across capacitor C will
appear at the output.

Mode III: Switch T is OFF, and current $IS = 0$ and $IP = 0$

The voltage across the switch T will also decrease because of Edc, so in
mode III, only Edc will appear and rest IP, IS and flux will be 0.

So, when the switch is ON, then a negative voltage appears at the output and when
the switch is OFF, then the positive voltage is at the output. Once current will be 0,
then the output voltage will also be 0. Since the secondary current will be 0, then the
output voltage will also be 0 and it will result in a discontinuity in the output.

The circuit diagram and waveform of the flyback power converter in the discontinuous mode are given in Figures 5.2 and 5.3, respectively.

5.2.2   FLYBACK POWER CONVERTER IN THE CONTINUOUS MODE

The continuous mode of the flyback converter has two modes named as Mode I and
Mode II only. The working of the flyback converter in the continuous mode is different in terms of the discontinuous mode because of the addition of one extra inductor
L, as shown in Figure 5.4. The working consists of two different modes.

Mode I: Switch T is ON, and current $IS = 0$

In the first mode when the switch is in the ON state, the primary winding
current IP having minimum current value initially increases gradually. The
secondary winding current is 0 and the voltage at the switch T is 0. The flux

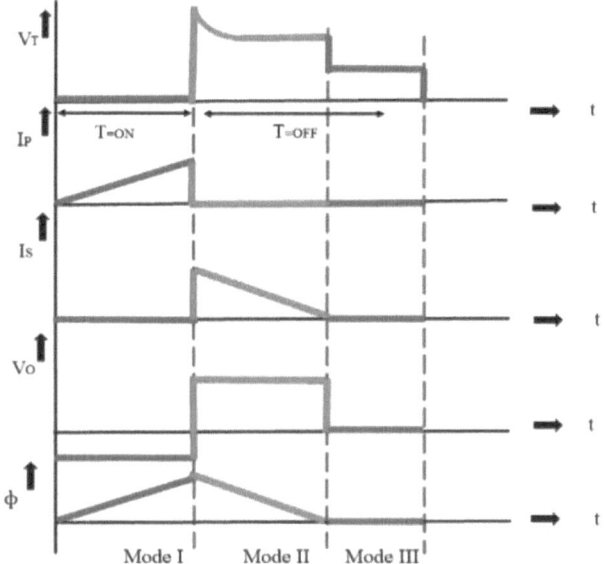

FIGURE 5.3 The waveform of the flyback power converter in the discontinuous mode.

having minimum current value initially also increases gradually and at the same time, the output is negative of *Edc*.

Mode II: Switch *T* is OFF, and current *IP* = 0

In mode II, the working is the same as that of discontinuous second mode except the current *IS* has some minimum value at first and goes on decreasing gradually and the declination in flux also takes place.

Thus, mode I and mode II keep on repeating and there is no mode III. The circuit diagram and waveform of the flyback power converter in the continuous mode are given in Figures 5.4 and 5.5, respectively.

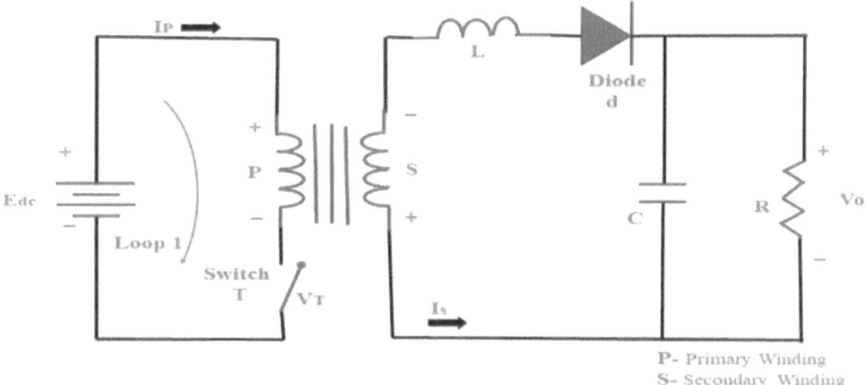

FIGURE 5.4 The circuit diagram of the flyback power converter in the continuous mode.

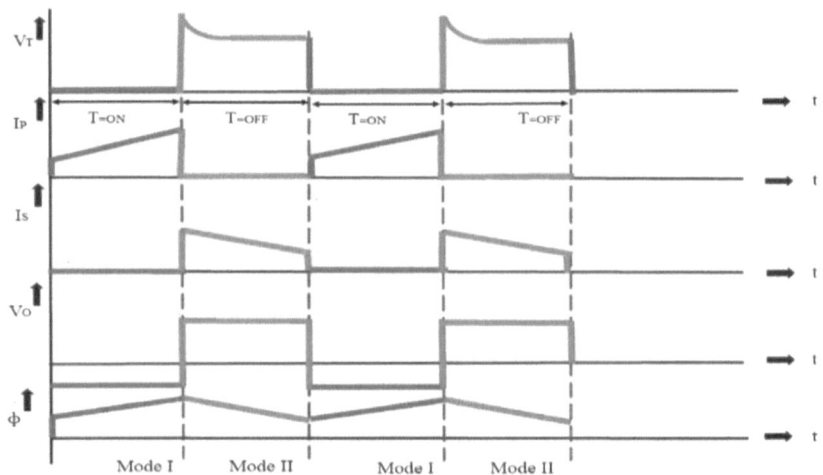

FIGURE 5.5 The waveform of the flyback power converter in the continuous mode.

5.3   BUCK-BOOST POWER CONVERTER

Buck-boost power converter falls into the category of a non-isolated DC/DC power converter. The working converter was based on three modes of operations. Though it is the same as that of flyback, just no barrier is present in the circuit.

Mode I: Switch *T* is ON
     In mode I operation, the switch *T* is ON where a transistor is acting as a switch. When the *T* is ON, the current is passed through Loop 1, during this stage diode *d* is in the OFF condition and at the same time, the inductor *L* is getting charged and voltage at the inductor *L* is V_{in}.

Mode II: Switch *T* is OFF
     In mode II operation, when the switch *T* is in the OFF state, the inductor *L* loses it energy and the polarity of the *L* is reversed to that of the polarity at the *L* in mode I. The inductor *L* energy is been released in Loop 2 where the current is passed to the capacitor C_2 as well as the load resistor *R* and during this period, the diode *d* is ON and the current *IL* is equal to the current *Id*. The capacitor C_2 is charged from the lower plate and diode *d* is ON until the energy is released in the inductor *L*.

Mode III: Switch *T* is OFF, and current *IL* = 0
     In mode III operation, when the inductor doesn't flow the current, then the current across the diode *d* is 0. The capacitor C_2 will flow current through Loop 3, i.e., in the output loop and the polarity at the output is the same as that of the C_2 means the output stage voltage is negative and that is the reason buck-boost power converter is called as inverting regulator. In this mode, voltage *VL* and current *IL*, *IT* and *Id* all are zero and this is termed as a discontinuous mode. To eliminate the discontinuous mode, the duty cycle *D* must be higher than the value 0.5.

The circuit diagram and waveform of the buck-boost power converter are given in Figures 5.6 and 5.7, respectively.

To know the working of the buck-boost power converter clearly, the solved equation helps to understand.

For mode I, When the switch T is ON,

$$T_{ON} = DT \tag{5.1}$$

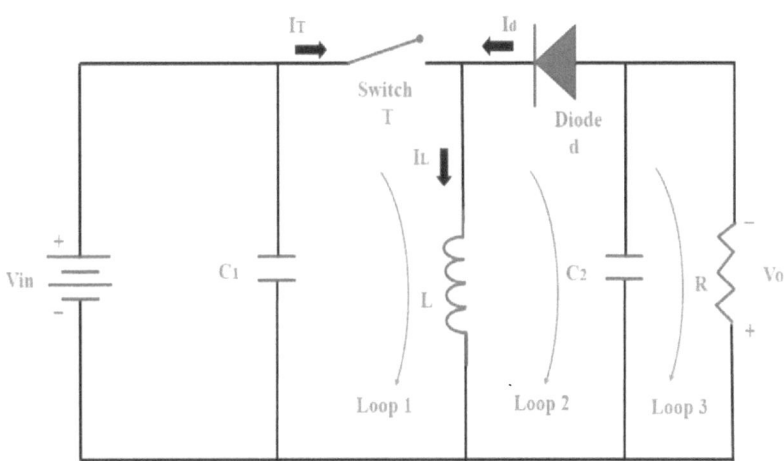

FIGURE 5.6 The circuit diagram of the buck-boost converter.

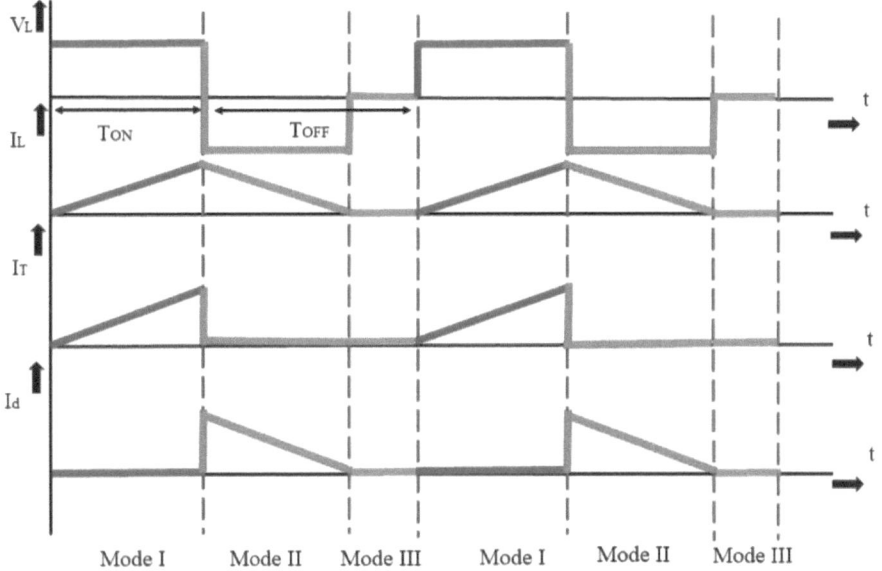

FIGURE 5.7 The waveform of the buck-boost power converter.

where,

T_{ON} = time when the switch is ON
D = duty cycle ($D = T_{ON}/T$, fswitching = $1/T$)
T = time period ($T = T_{ON} + T_{OFF}$)

Now voltage at inductor L, when the switch is in the ON state,

$$(VL)_{ON} = V_{in} \tag{5.2}$$

For mode II, when the switch T is OFF,

$$T_{OFF} = (1 - D)T \tag{5.3}$$

where,

T_{OFF} = time when the switch is OFF

Now voltage across inductor L, when the switch is OFF,

$$(VL)_{OFF} = -V_o \tag{5.4}$$

Using the volt-sec. across the inductor,

$$(VL)_{ON} T_{ON} + (VL)_{OFF} T_{OFF} = 0 \tag{5.5}$$

$$(V_{in})DT + (-V_o)(1 - D)T = 0 \tag{5.6}$$

$$V_{in}D = V_o(1 - D) \tag{5.7}$$

$$V_o = V_{in}\left[D/1 - D\right] \tag{5.8}$$

The duty cycle varies in the values 0 and 1. From Equation 5.8, when D lower than 0.5 is given, the circuit will work as the buck converter and when D is higher than 0.5, the voltage at the output is larger than the voltage at its input, then the circuit will function as the boost power converter.

5.4   STEPUP DC/DC CONVERTERS

Renewable power generation systems are more attracted by the government and industries because of their advantages like pollution-free and availability being free of cost. The DC/DC converters place a vital role in the integration of PV and wind systems [8]. The researchers are very interested in designing various DC/DC for non-conventional energy source applications. This section reviews various DC converter topologies which can provide better efficiency, good conversion ratio, less operating losses and low conduction losses.

5.4.1 Conventional Boost Converter

Figure 5.8 shows the design of a conventional boost converter which gives high voltage gain if its duty cycle reached 1. High conduction losses and reverse recovery issues are caused due to the increase in the value of the duty cycle. Due to this, the overall efficiency of the system decreases. The limitations of this converter are current ripple, voltage stresses on transistors and increased recovery losses. Three-level converters provide double voltage gains when compared with the two-level inverter. It minimizes the voltage stresses, recovery losses and improves efficiency [9].

5.4.2 Interleaved Stepup Converter

A general stepup converter uses four interleaved boost converter configurations. This structure distributes the input current eventually; hence, it reduces the switching losses and output voltage ripple [10]. An interleaved boost converter which reduces the input current ripple is the use of isolation transformer [3]. This structure uses the AC link. For improving the voltage gain, it uses the voltage multiplier units, as depicted in Figure 5.9. It helps for producing high voltage gain [7]. High voltage gain flyback converter uses the LC filters for filtering the harmonics which minimizes the voltage stresses of the switches [1].

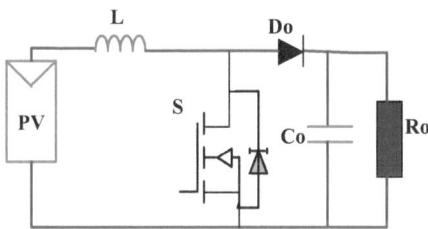

FIGURE 5.8   Single switch boost converter.

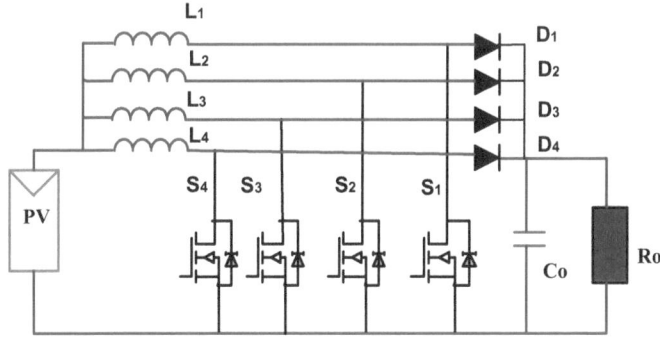

FIGURE 5.9   Interleaved boost converter.

5.4.3 Soft Switching High Stepup DC/DC Converter

Boost converters consist of switches and passive components. This converter uses soft switching techniques such as Zero Voltage Switching (ZVS) and Zero Current Switching (ZCS) for operating the switches. Due to the use of soft switching techniques, it can effectively eliminate the switching losses. The soft switching-based high-performance boost converter is shown in Figure 5.10.

5.4.4 Coupled Inductor-Based Boost Converter

The model generally acts like a transformer in Figure 5.11. It has better efficiency and good voltage gain. This structure can use voltage doubler circuits and active clamp circuits. To minimize the voltage stresses on the switches, primary passive clamp circuits are placed at the primary side [4]. Boost converter with a voltage doubler feature is utilized to boost the voltages and to reduce the voltage stresses. Soft switching techniques are employed along with voltage doubler circuits to achieve high voltage gain and reduce stresses on switches by raising the turn's ratio of the system [6].

5.4.5 Interleaved with Voltage Multiplier Cell

The developed system is shown in Figure 5.12, which uses the current division technique. In classical interleaved boost converters, MOSFET and voltage multiplier

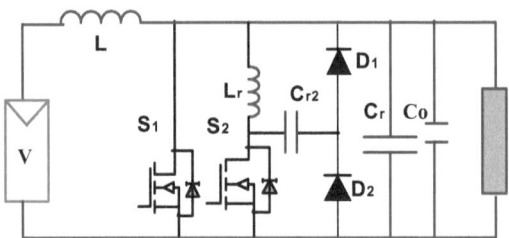

FIGURE 5.10   Soft switching ZCZVS boost converter.

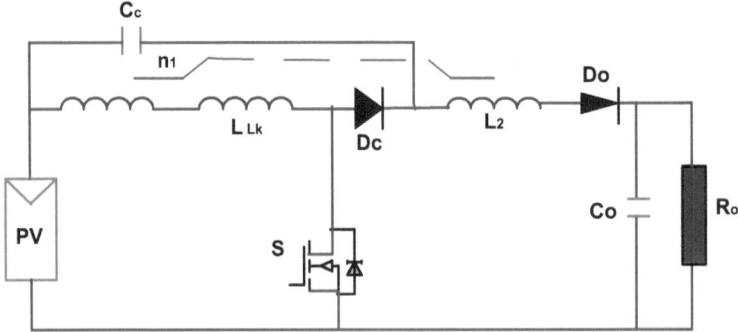

FIGURE 5.11   Coupled inductor-based boost converter.

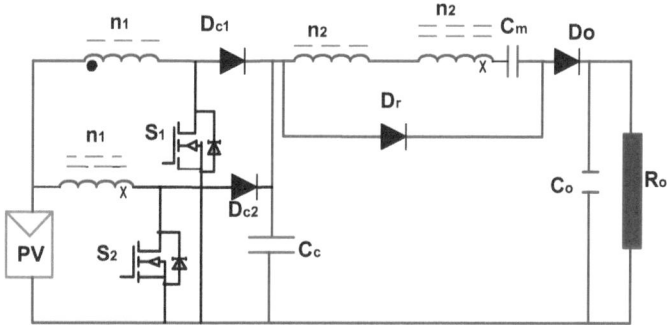

FIGURE 5.12 Interleaved boost converter with the voltage multiplier cell.

circuits are used for successfully reducing the harmonics in output voltage, turn-on losses and electromagnetic interference [2]. For high power applications, better performance and smaller size converters are required. This can be achieved by the utilization of clamp circuits and soft switching. It reduces the harmonics, voltage stresses and passive element size [7].

5.4.6 ISOLATED BOOST CONVERTER WITH COUPLED INVERTER

This structure uses a single magnetic circuit instead of individual magnetic circuits. Due to the assembling of all magnetic circuits into a single magnetic circuit, the size of the capacitor and inductor is decreased, as shown in Figure 5.13. It is useful for a wide range of power applications. The disadvantage of this structure is it uses hard switching [5]. The structure with soft switching is used for low power applications that have low stress on switches and low harmonics [1].

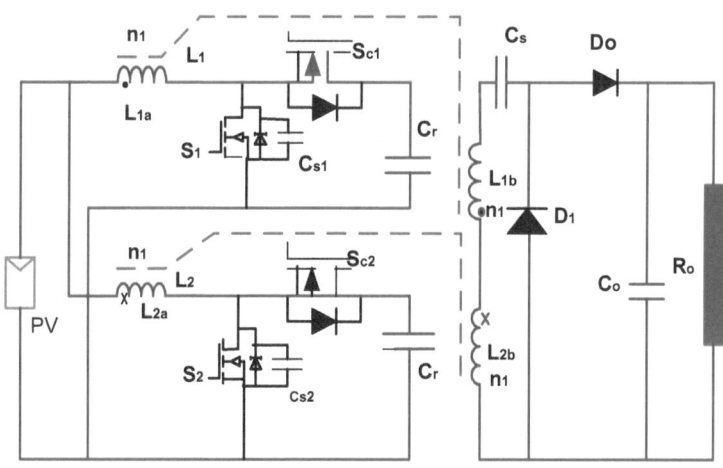

FIGURE 5.13 Isolated boost converter with the coupled inductor.

5.4.7 NON-INDUCTIVE BOOST CONVERTER

This is a back to back multi-level DC/DC boost converter, which has six times of voltage conversion ratio. This structure performs buck and boost operations. It works as a buck converter when the power flow is reversed. It can operate at high temperatures because of the absence of inductor elements. It has advantages of high efficiency and good voltage regulation [2]. Capacitor coupled boost converter will operate at very high frequencies. It can also have bidirectional power flow and low voltage drop, which is depicted in Figure 5.14.

5.4.8 BASIC ISOLATED BOOST CONVERTER TOPOLOGY

There are many converters such as flyback converter, zeta converter, push–pull, cuk and two inductor boost converter has provided more output gain but failed in providing higher efficiency. The active clamp boost converter is shown in Figure 5.13. It has good output voltage, efficiency and synchronization with the DC link [3]. The resonant converter shown in Figure 5.14 uses the snubber configuration and voltage doublers for reducing on/off losses. The LC resonant converter output diode eliminates the reverse recovery issues, hence it is useful to use at various PV applications (Figures 5.15 and 5.16) [6].

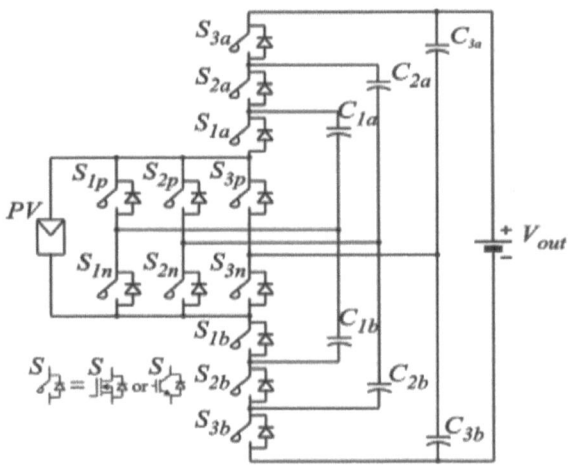

FIGURE 5.14 Multilevel switched capacitor boost converter.

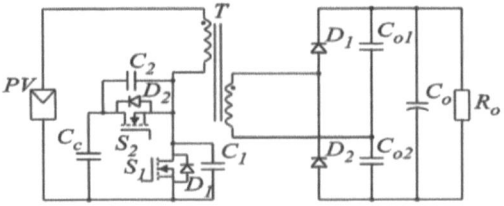

FIGURE 5.15 Active clamp boost converter.

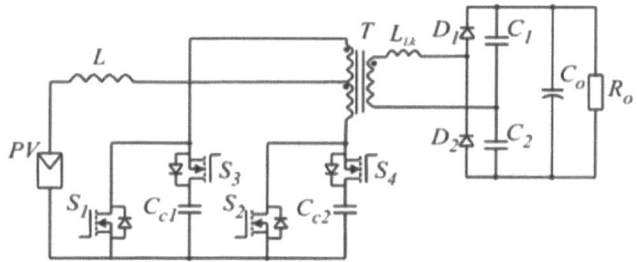

FIGURE 5.16   Push–pull resonant boost converter.

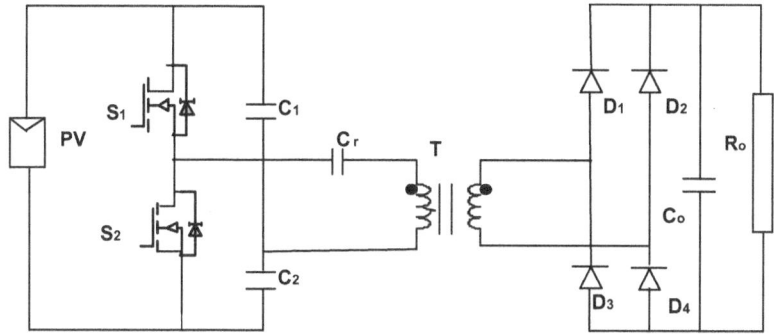

FIGURE 5.17   Resonant half bridge boost converter.

5.4.9   HALF BRIDGE RESONANT CONVERTER

For high output voltage, generally current-fed half bridge resonant converter is utilized, which is shown in Figure 5.17. In this structure, due to the transformer turn
off losses, the efficiency reduced [10]. To overcome this problem, the transformer
leg diode is replaced with the leg capacitor. The leg capacitor effectively reduces the
turns ratio and cost [9].

5.4.10   NON ISOLATED CONVERTER WITH INBUILT TRANSFORMER

High voltage gain of conventional converters is controlled by passive elements.
Lowering the duty cycle reduces the efficiency and gain [10]. To degrade this problem in traditional converters, a novel technique is developed with a single switch
three diode topology which is depicted in Figure 5.18. This has the advantage of a
good power factor correction circuit and smaller in weight.

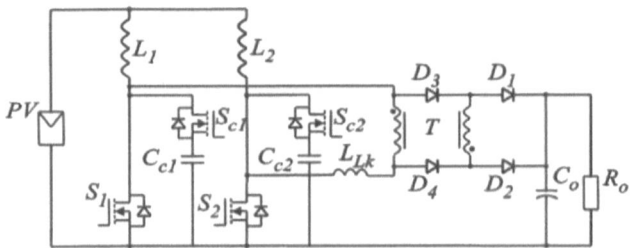

FIGURE 5.18 Non-isolated high stepup converter.

TABLE 5.1

Comparison between the Parameters of the Designed Power Converter

Parameter	Flyback Converter (Discontinuous)	Flyback Converter (Continuous)	Buck-Boost Converter (D Less Than 0.5)	Buck-Boost Converter (D Higher Than 0.5)
Input voltage (v)	220	220	220	220
Capacitor (μF)	47	47	47	47
Inductor (mH)	1	1	1	1
Resistor (Ohms)	10	10	10	10
Duty cycle	0.5	0.5	0.3	0.7
Output voltage (v)	54.16	54.16	−93.54	−482.2
Output current (amp)	5.416	5.416	−9.354	−48.22

5.5 RESULT

The flyback power converter with continuous and discontinuous as well as buck-boost power converter with duty cycle higher and lower than 0.5 were designed and simulated in MATLAB Simulink. The comparison between them is tabulated in Table 5.1 based on the voltage and current at the output side.

5.6 CONCLUSIONS

From Table 5.1, the conclusion was drawn that when the same values of the voltage at the input are taken, as well as the same value of capacitance, resistance, inductance having duty cycle 0.5 are taken for the flyback converter, the value of current and voltage at the output is same. 0.3 and 0.7 values of D are taken for the buck-boost power converter. The buck-boost converter acting as a boost power converter for given D is greater than 0.5, and for D less than 0.5, the buck-boost converter is acting like a buck power converter. The current and voltage are the same for the flyback converter in both continuous and discontinuous modes at its output. The one thing again noticed is that the current and voltage are negative for the buck-boost power converter at the output. From the above result, it is clearly shown that the buck-boost power converter has maximum voltage and current at the output as compared to the flyback converter, but the output is found negative.

REFERENCES

1. S. Malapur, M.B. Shruthi, N. Bodravara, Davanagere, "Comparison of Isolated and Non-Isolated Bidirectional DC–DC Converter Fed PMDC Motor", *International Journal for Research in Applied Science & Engineering Technology (IJRASET)*, 3, Issue IV, April 2015, ISSN: 2321-9653.
2. R. Zgheib, I. Kamwa, K. Al-Haddad, "Comparison between isolated and non-isolated DC/DC converters for bidirectional EV charges", *2017 IEEE International Conference on Industrial Technology (ICIT)*, March 2017.
3. K.V.G. Raghavendra, K. Zeb, A. Muthusamy, T.N.V. Krishna, S.V.S.V. Prabhu Deva Kumar, D-H. Kim, M-S. Kim, H-G. Cho, H-J. Kim, "A Comprehensive Review of DC–DC Converter Topologies and Modulation Strategies with Recent Advances in Solar Photovoltaic Systems", *MDPI, Electronics*, 9, p. 31, 2020.
4. E. Baghaz, S. Bounouar, R. Bendaoud, B. Zohal, N.K. M'Sirdi, A. Naamane, N. Benaya, N. El Akchioui, M. Benhmida, "Study and Design of a Full Bridge DC/DC Power Converter", *Universal Journal of Electrical and Electronic Engineering*, 6, Issue 2, pp. 31–45, 2019.
5. R. Erickson, D. Maksimovic, *Fundamentals of Power Electronics*, 2nd ed., Chapter 19, Springer, 2001.
6. J. Falin, "Designing DC/DC Converters Based on ZETA Topology", *Analog Applications Journal*, Texas Instruments Incorporated, pp. 16–21, 2010.
7. N. Kasundra, A. Kumar, "Design and Simulation of Flyback Converter in MATLAB using PID Controller", *International Journal of Advanced Research in Electrical, Electronics and Instrumentation Engineering*, 5, Issue 2, pp. 960–964, February 2016.
8. N. Mohan, T.M. Undeland, W.P. Robbins, *Power Electronics: Converters, Applications and Design*, ISBN: 978-0-471-22693-2.
9. M. H. Rashid, "Power Electronics: Circuits, devices and Applications" ISBN 978–0131011403
10. B.S. Goud, B.L. Rao, C.R. Reddy, An Intelligent Technique for Optimal Power Quality Reinforcement in a Grid-Connected HRES System: EVORFA TECHNIQUE, *International Journal of Numerical Modelling*, 34, p. e2833, 2021, doi: 10.1002/jnm.2833.

6 Controller for a Remotely Located PV/Wind Hybrid Energy System Implemented via Internet of Things

Sonalika Kishore Kumar Awasthi
Electronics and Communication Engineering,
Vellore Institute of Technology

Meher Jangra
Vellore Institute of Technology

Sasikumar Periyasamy
Vellore Institute of Technology

CONTENTS

DOI: 10.1201/9781003143802-6

NOMENCLATURE

PV	Photovoltaic
WE	Wind energy
DG	Diesel generator
RES	Renewable energy systems
BES	Battery energy storage
HRES	Hybrid renewable energy system
GPIO	General purpose input/output
SSH	Secure shell
E_T	Total generated power
E_S	Power generated by solar panels
E_W	Power generated by WE system
E_L	Load power
E_B	Power consumed by LED bulb
E_F	Power consumed by fan
V_{LAB}	Voltage of the lead acid battery
WT	Wind turbine
DL	Dump load
PWM	Pulse width modulation
MPPT	Maximum power point tracking

6.1   INTRODUCTION

Realization of the fact that our unhealthy consumption of nonrenewable energy sources has been the leading cause of weather changes has driven scientists and technology giants toward solar and wind energy. Wind energy systems independently are not always viable because of sites with low wind speeds and are hence solar energy is less unpredictable than wind. There is a statistical rise in the demand of renewable energy in India and such patterns are also expected in the future (Figures 6.1 and 6.2). Even though India began its steps for integrating renewable energy systems (RES) in the late 80s, it has made high-scale advancements given the enormous size of the country. Wind power has clearly proven to be the most fruitful renewable energy (Ritchie and Roser 2020). By 2022, India shall be yielding 175 GW of power only on the basis of renewable energy sources from which 60 GW shall be contributed by wind energy. Hence, the combined utilization of these two is becoming increasingly attractive. Hybrid renewable energy systems (HRES) are hence very popularly applicable to generate energy for remote areas. It is also vital to perfect these hybrid energy systems for effective implementation. Given the developments

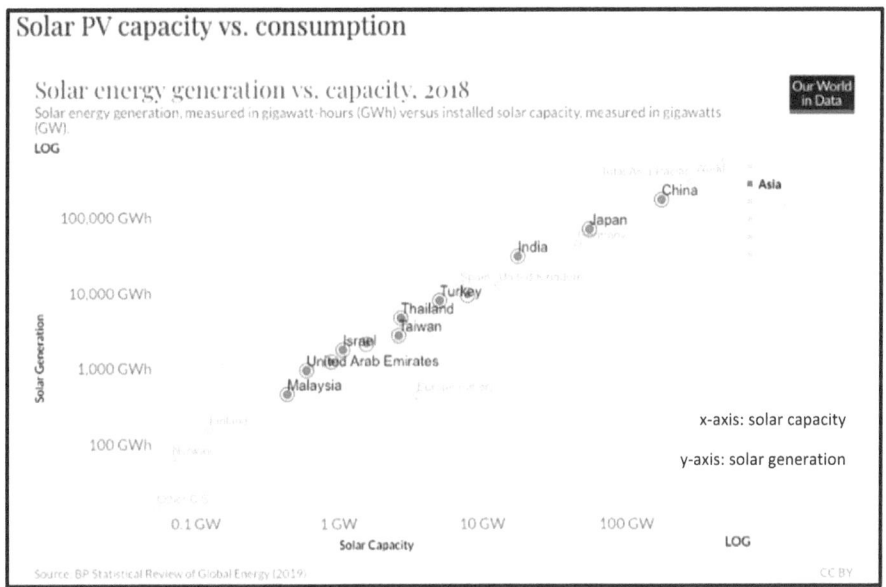

FIGURE 6.1   Solar PV capacity vs. consumption.

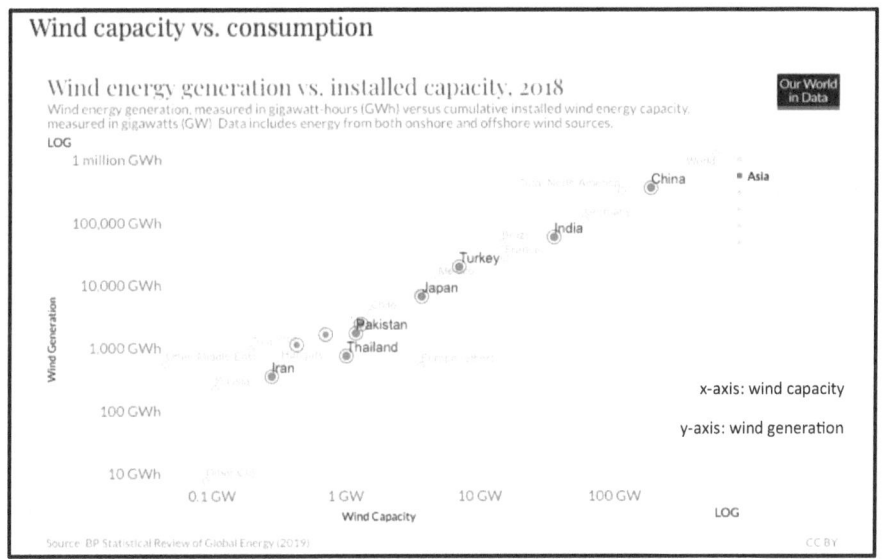

FIGURE 6.2   WE generation vs installed capacity.

in technology, various applications such as sensors, transmitting, distributing and flexibility of availability of electricity, it is very feasible and easy to monitor and control the source of energy powering this system which is constructed for this project. Due to the inherent properties of renewable energy resources, it is quite difficult to implement an efficient integration of a system which caters to the power consumption needs of remote areas. PV or WE singlehandedly cannot meet the demands of the DC loads in remote areas. This makes the installation of HRESs in such areas very vital for thriving commercial and domestic purposes. Need of HRESs is also important in areas which cannot support the construction of grid extensions. There is also a clear lack of efficient systems for controlling HRES. Hence for the urgent requirement of their effective implementation, establishment of a system which can control them is necessary. Because of the benchmarks set by numerous universal associations, a few nations have been urged to build micro-grid power generation facilities to control the remote areas that can't be associated with the network because of financial and specialized constraints. By utilizing the economies of scale, it is prudent to use RESs to satisfy the heap need at the rural communities than to burn through a great many dollars for augmentation of transmission and distribution lines to such regions with no money-related advantages (Tan 2019).

6.2 FEASIBILITY ANALYSIS OF HYBRID SYSTEMS

Wind, geothermal, solar and tri-generation systems are standalone and have proven to be less efficient than higher yielding HRESs in terms of environmental and economic returns (Gupta et al. 2006; Venkatraman et al. 2016). Storage of energy can possibly build the adaptability of the framework since it loosens up the worldview that power age must match the addition of load and losses everytime. Additionally, as the energy from inexhaustible sources increases, progressively controllable assets must be utilized so as to adapt to its low consistency and high fluctuation. Still, experience and developing exploration have demonstrated that PV and WE can supplement each other up to a specific level as far as time of accessibility and pinnacle of power are concerned. This complementarity gets valuable in areas where it is difficult to attain other power resources to smoothen out the variances in PV energy. This has led numerous nonrenewable energy to tap into the PV and wind hybrid systems to generate energy and this practice is gaining even more popularity because of the parity of wind and solar power generation cost. For these reasons, renewable energy is more penetrable in energy storage than ever (Chaib et al. 2016). Dhage et al. (2018) and Das et al. (2005) propose modeling of PV + wind + fuel cell HRES.

In worst environmental conditions, the PV and wind energies shall fail to provide but the fuel cell will be effective to provide 10 kW of power. They put forward a non-complex and feasible method for controlling with which a DC–DC converter is used for maximum power point monitoring which leads to maximum energy generation from the wind turbine and solar panels.

Another study (Chaib et al. 2016) is in line with this chapter which aimed for the following: to manage the extracted power, it is accomplished through two connector switches in the DC output of the two sources. Every subsystem switch is controlled so as to power either a dump load or the customers' needs through an inverter. The goal is to acquire a satisfactory framework response and to make it increasingly efficient. Moving forward with a study(Venkatraman et al. 2016) which proposes a novel solution of controlling the HRES using a FPGA, IOT is used to control the energy supply, i.e., RNE or NRE of a house through a secure website when the grid supply is off/not present.

6.2.1 PROPOSED SYSTEM

In the proposed system (Figure 6.3), the Raspberry Pi 3 model B+ is utilized to receive and transmit data (control signals and power management) wirelessly, which is dealt with on the NodeRed platform designed to control the system. The main criteria are controlling what powers the loads out of the two energy sources, i.e., HRES and nonrenewable energy (the grid or DG) without any inconvenience through a website using Raspberry Pi 3 model B+.

There will be two types of loads dealt with: critical and noncritical. The Raspberry pi will enable manual switching between power sources and also keep in the user's POV the voltages and automatically switch if needed. This empowers the client to have adaptable control instruments remotely through a made sure about web association. This framework encourages the client to control the sources of energy, physically and remotely utilizing a smart mobile phone or PC. This framework is proficient, less expensive and adaptable in operation.

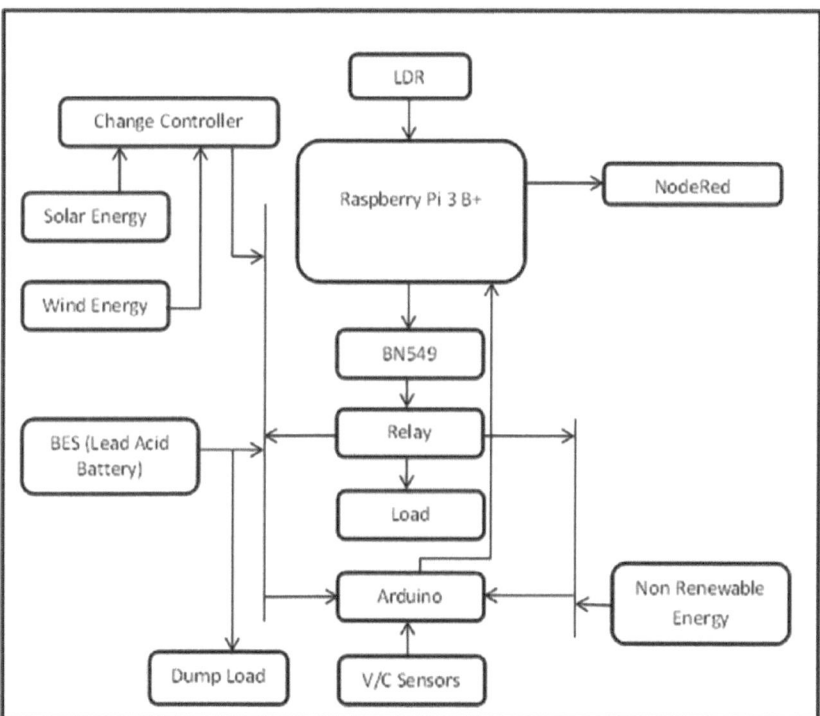

FIGURE 6.3 Block diagram of the proposed system.

6.3 SYSTEM STRUCTURE AND MODULES

The system proposed in this chapter consists of a raspberry pi 3 model B+ as its core component which functions as a microcontroller unit required in order to carry out the processing of the input data. Sensors like INA 219 and LDR have been interfaced with the raspberry pi. This setup is further connected via WiFi to the internet in order to facilitate the transfer of data on the NodeRed server. The NodeRed platform was accessed via SSH, and the dashboard displays the status of the energy system. Along with a manual switch, the system also contains nodes for automatic switching when the sensors sense favorable conditions for charging the renewable system BES.

6.3.1 RASPBERRY PI MODEL B

The Raspberry Pi is a mini computer that uses a standard keyboard and mouse and functions after being plugged into a computer monitor or TV. Besides being a computer that is low cost and runs Linux, it also allows the user to control the components involved for successful physical computation and helps in exploring the opportunities of Internet of Things (IoT). This super useful device helps people across all ages to learn programming using languages like Python and Scratch, thereby enabling them to explore computing. Raspberry Pi finds its use in a plethora of technical projects spread across multiple areas like weather stations, music machines, parent detectors, and tweeting bird houses having infra-red cameras by virtue of its ability to establish a connection with the outside world.

6.3.2 SOLAR ENERGY SYSTEM

A photovoltaic module consisting of an installed framework of an assembly of mounted photovoltaic cells is also referred to as a solar panel. These cells run on sunlight as its source of energy facilitating the generation of direct current electricity. An array consists of a collection of panels and such a PV panel consists of a system of modules. The supply of solar electricity to all the electrical equipment is carried through these arrays of a photovoltaic system.

6.3.3 SOLAR CHARGE CONTROLLER IMPLEMENTING MPPT

The task of power management is to direct the power from the solar array to the battery bank, which is taken care of by a solar charge controller (Figure 6.4). This component keeps in check the charging of the deep cycle batteries making sure that during the day they are not overcharged. It also prevents the drainage of the battery by making sure that overnight the power doesn't run backward to the solar panels. Even if some of the charge controllers are equipped with carrying out extra functions like load and lighting control, still their primary job will be power management. MPPT and PWM are the two different technologies that a solar charge controller is available in. An MPPT, i.e., maximum power point tracking solar charge controller measures the Vmp panel's voltage and down converts the PV voltage to the battery voltage. In order to match the battery bank, we raise the current; because the input

FIGURE 6.4   Solar charge controller.

power to the solar charge controller is the same as the power output of the solar charge controller, thereby utilizing more of the panel's available power. A solar array of a higher voltage can be used instead of a battery. For instance, the 60 cell nominal 20 V grid-tie solar panels being more readily available can be used. A 12 V battery bank can be charged using a 20 V solar panel, two of those in series can charge up to 24 V battery bank and three of those in series can charge a battery bank of 48 V. The off-grid solar system can use this, thereby utilizing this wide range of solar panels.

6.3.4   SWITCHING MECHANISM

- BC549 NPN Transistor

  NPN transistors have a p-type material placed between two n-type materials. It produces strong amplified signals at the collector end after amplifying the weak signals entering into the base. The transistor is where the current constitutes because the path of electron movement is from the emitter to collector region in NPN transistors. Electrons which have high mobility compared to holes are their majority charge carriers and thus such types of transistors are mostly used in the circuit. BC547, BC548, and BC549 belong to a group of general purpose NPN semiconductors of which the NPN transistor, BC549C is a part of. They find use in small signal directly coupled preamplifiers and audio frequency. The general purpose range of transistors being equivalent and similar to each other can serve as a good substitute for applications like basic noncritical audio applications (Figure 6.5).

- SRD-05VDC-SL-C

  The switches that open and close circuits electronically or electro mechanically are called Relays. An electrical circuit is controlled by opening and contacts in another circuit when relays are involved. When the relay is not energized and there is an open contact, then the relay contact is normally open (NO). When the relay is not energized and there is a closed contact, the relay contact is normally closed (NC). These states in both the cases can be changed by applying electrical current. Except for devices

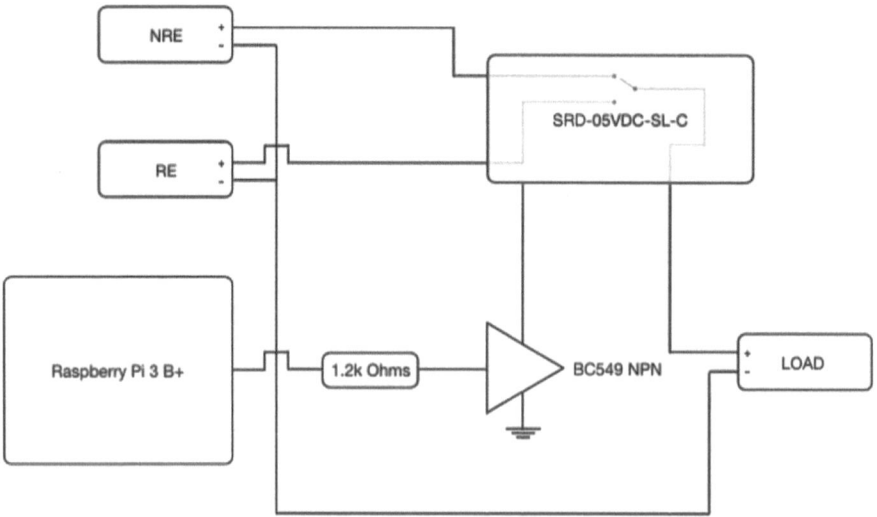

FIGURE 6.5   Switching mechanism.

drawing low amps like Solenoids and small motors, relays do not generally control power heavy devices and are usually used in a control circuit to switch smaller currents. It is possible however that a small voltage applied to a relay coil can lead to a large voltage switch triggered by the contacts, thereby having an amplifying effect and thus controlling larger voltages and amperes.

6.3.5   WIND ENERGY SYSTEM

The wind energy system is emulated via two DC motors. Various systems (Dhage et al. 2018; Das et al. 2018; Venkatraman et al. 2016) built used HOMER to simulate the solutions they designed, while this chapter used a generator and wind turbine to construct the wind energy system.

6.3.6   POWER MANAGEMENT

The BES for the RES consists of two 12 V lead acid batteries. The power across every component shall be recorded and monitored at all times and sent to the NODERed platform. It shall be implemented using the INA219. As an easy solution to power monitoring and management solution, it is highly effective. INA219B also saves the effort of using separate sensors for sensing current and voltage. It uses the TWI/I2C of the host to digitally send the data. The sensor can also measure a high side voltage and current of 26 V and 3.2 A, respectively. The precision current amplifier gauges the voltage over the 0.1 Ω, 1% sense resistor to decide the current in the loop. The joined voltage and current estimations permit capacity to be determined, making the INA219 extraordinary for following battery life or solar panel boards. (Power = Voltage * Current). The implementation shall be done via connecting an INA219B to the Raspberry Pi which shall be used to keep in check the power management of the solar power. Two Arduino Uno boards with voltage and current sensors shall be connected to the Raspberry pi via USB cables as external serial inputs.

6.3.7 NODERED

NodeRed is a visual programming platform for connecting hardware devices, APIs and services managed on the internet in new and easy ways. It has a browser-based visual editor that makes it easy to wire together nodes to construct flows. The palette on the platform has a wide range of nodes which can also be programmed and morphed into brand new nodes with alternate functions. NODERed hence is a very functional platform to be integrated with raspberry pi with easy access and great productivity and ease. It enables applications like these to be deployed with a single click.

6.4 MATHEMATICAL MODEL

The power calculations are carried out as follows.

$$E_T = E_S + E_W \tag{6.1}$$

$$E_L = E_B + E_F \tag{6.2}$$

6.4.1 LOGIC OF LOAD CONTROL

Considering LED bulb as the critical load and fan as the noncritical load.

$$\text{If} \quad E_L \geq E_T \Rightarrow \text{Switch OFF fan}$$

Thus, the power is used up only to light up the LED bulb, our critical load. Now, if E_T increases and if

$$E_T \geq E_L \Rightarrow \text{Switch ON fan}$$

6.4.2 LOGIC OF SOURCE CONTROL

If $E_T < E_L$ and fan is off and $V_{LAB} < 19\,V \Rightarrow$ Switch ON nonrenewable energy and put lead acid battery to charge.

And if, $V_{LAB} > 30\,V \Rightarrow$ Switch OFF nonrenewable energy and put lead acid battery to discharge.

6.4.3 LOGIC OF SWITCHING DL AND WT

If $E_T \geq E_L$ and $V_{LAB} > 30\,V \Rightarrow$ Switch ON DL and start T_{DL}
If $E_T \geq E_L$ and $V_{LAB} > 30\,V$ and $T_{DL} > 10\,\text{min} \Rightarrow$ Switch OFF WT and DL.

6.5 CONTROL AND MONITORING IMPLEMENTATION

6.5.1 Renewable vs Nonrenewable

The LDR and prospected wind speedometer readings are read on the Raspberry pi 3 Model B+ and sent to the NodeRed web platform where it is used to determine whether the conditions are favorable for the renewable system to yield power. If they are, the system switches to the nonrenewable energy (DG) to power the load. This is brought to effect via a relay. This relay is triggered using the NodeRed dashboard. The nodes and Rpi GPIO setting are shown in Figure 6.6.

6.5.2 Renewable BES Charging vs Discharging

The INA219 employed to keep in check the current, voltage and hence power at various terminals of the system is fetched via NodeRed and displayed to the dashboard. When the nonrenewable energy source is put in use and switched on, the same switch automatically triggers the relay connected between the solar charge controller (Figures 6.7 and 6.8), the load and the lead acid battery (i.e., the renewable BES). This triggers the BES to automatically shift to the charging mode.

6.6 RESULTS

The voltage and current values obtained from the serial port of the Raspberry pi via the Arduino Uno are run through parsing nodes to separate the serially received values. Via further conditioning using function nodes, the powers are calculated. Furthermore, the comparisons for load and power management are implemented via function nodes. Dashboard is constructed with dashboard palette nodes such as gauges, charts, switches and input texts (Figures 6.9–6.11).

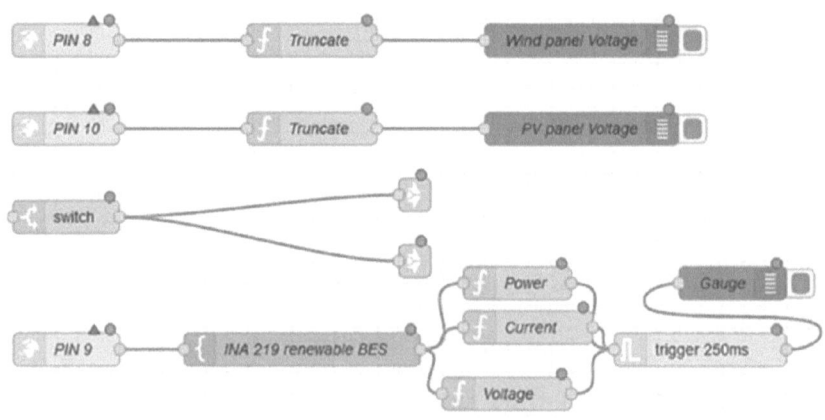

FIGURE 6.6   NODERed flow (switching).

FIGURE 6.7   Solar charge controller switching to the discharge mode.

FIGURE 6.8   Solar charge controller switching to the charging mode.

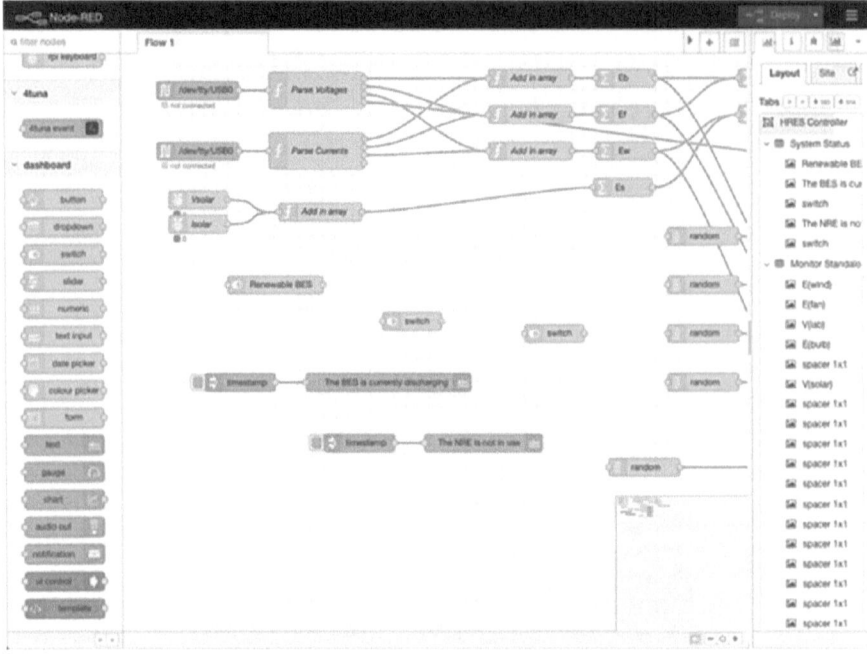

FIGURE 6.9   Complete NODERed flow.

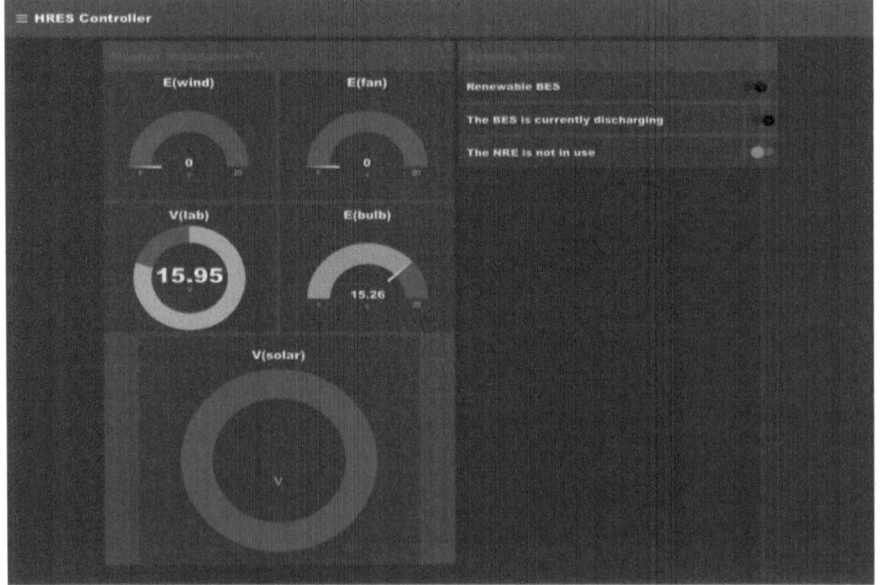

FIGURE 6.10   HRES dashboard.

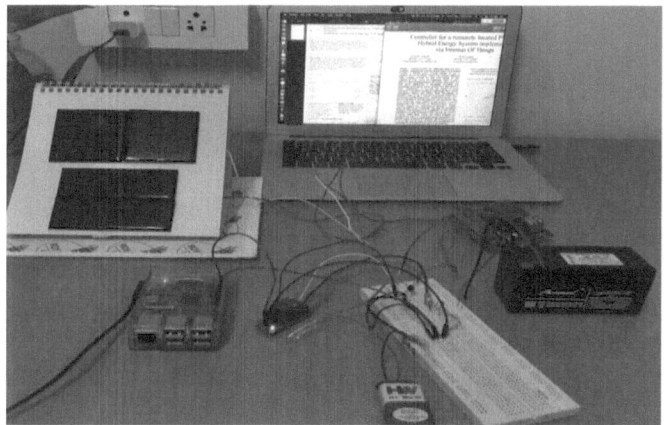

FIGURE 6.11 Project setup.

6.7 CONCLUSIONS AND FUTURE WORK

Driven by the goal of exploring the attractive avenue of hybrid energy systems and efficient management of its extensive potential, the proposed prototype was successfully built. The electrical data was sent and received via Rpi and further dealt on NodeRed for controlling purposes. A combination of solar energy array and WE system consisting of a generator and wind turbine was the renewable energy sources. The BES for the RES consisted of two 12 V lead acid batteries. INA219 was connected to the RPi in order to check the power management of the solar panel. Logic calculations to channelize the generated power were carried out considering LED bulb and fan as the critical and noncritical load, respectively, and required power switching decisions were made. Comparisons for load and power management were implemented via function nodes on NodeRed. At the NodeRed server, the readings sent by RPi of the LDR and prospected wind speedometer were used in deciding to switch the system to nonrenewable energy (DG) to power the load provided the conditions were favorable for the renewable system to yield power. This was brought to effect via a relay triggered using the NodeRed dashboard which consisted of dashboard palette nodes like gauges, charts, switches and input texts. When the nonrenewable energy source was put in use and switched on, the same switch automatically triggered the relay connected between the solar charge controller, the load and the lead acid battery triggering the BES to automatically shift to the charging mode. For future work, we can transform this system into a real-world working system involving a standalone AC micro-grid made via a WE generator and a PV panel as varied generators. Using enough electronic energy converters at the output of the RNE sources, a DC bus can be merged through a DC–DC converter for storing energy. MPPT controllers would ensure maximum power extraction of RNE sources and further pass it to the DC line. A three phase inverter can be used to extract as per phase AC voltage which would act as a micro-grid which will be a standalone system for powering a collection of critical and noncritical loads.

REFERENCES

Chaib A., Achour D., Kesraoui M. Control of a solar PV/wind hybrid energy system. *Energy Procedia*, 95: 89–97, 2016. doi: 10.1016/j.egypro.2016.09.028

Das D., Esmaili R., DaveNichols L.X. An optimal design of a grid connected hybrid wind/photovoltaic/fuel cell system for distributed energy production. *IEEE*, 23(5): 2499–2505, 2005.

Das G., De M., Mandal K.K., Mandal, S. Techno economic feasibility analysis of hybrid energy system. 2018 Emerging Trends in Electronic Devices and Computational Techniques (EDCT), 2018. doi: 10.1109/edct.2018.8405060

Dhage S., Pranjale M., Jambhulkar S., Warambhe N., Review on grid connected solar wind hybrid power based IOT system. *International Research Journal of Engineering and Technology (IRJET)*, 5(02), 1346–1348, Feb 2018.

Gupta A., Saini R.P., Sharma M.P. A Hybrid AC/DC Microgrid and Its Coordination Control. *IEEE Transactions on Smart Grid*, 2(2), 278–286, 2006.

Pradhan S., Singh B., Panigrahi B.K., Murshid S., A composite sliding mode controller for wind power extraction in remotely located solar PV–wind hybrid system. *IEEE Transactions on Industrial Electronics*, 66(7), 5321–5331, July 2019.

Ritchie H., Roser M. Renewable energy. Published online at OurWorldInData.org. Retrieved from: https://ourworldindata.org/renewable-energy [Online Resource], 2020.

Tan Y. Enhanced frequency regulation using multilevel energy storage in remote area power supply systems. *IEEE Transactions on Power Systems*, 34(1), 163–170, Jan. 2019.

Venkatraman K., Dastagiri Reddy B., Selvan M.P., Moorthi S., Kumaresan N. Online condition monitoring and power management system for standalone micro-grid using FPGAs. *IET Generation, Transmission & Distribution*, 10(15), 11–17, 2016.

Zhu H., Zhang D., Athab H.S., Wu B. PV Isolated Three-Port Converter and Energy-Balancing Control Method for PV-Battery Power Supply Applications. *Transactions on Industrial Electronics*, 62(6), 3595–3606, June 2015.

7 Facile Hydrothermal Synthesis of CuS/MnS Composite Material Improved the Power Conversion Efficiency of Quantum-Dot Sensitized Solar Cells

Ikkurthi Kanaka Durga
Pusan National University

Suresh Alapati and Srinivasa Rao Sunkara
KyungSung University

CONTENTS

7.1 INTRODUCTION

Increasing energy demands, global population and the corresponding energy need to be addressed productively using renewable energy sources. In the past few years, the global climate is threatened by cohort of power and an anthropogenic change without discharge of poisonous gases is essential in existing scenario [1,2].

DOI: 10.1201/9781003143802-7

As we know, the green technologies have discovered the use of water and air for power generation. Due to high consumption of energy, subsequent advances in the growing technology as well as rapid increase in the global population such as green technologies are inadequate to accomplish the above requirement. The exorbitant production and/or fabrication cost of thin film solar cells irrespective of high-power conversion efficiency (PCE) led to the development of dye-sensitized solar cells (DSSCs) as an easy fabrication and low-cost alternative by O'Regan and Gratzel in 1991 [3–5]. The emergence of DSSCs has created a new tool for their effectiveness in converting incident photons to electrical current and is of great interest due to their low-to-medium purity materials than can be used. However, the photo degradation and instability of DSSC still make them inferior to conventional Si and thin film solar cells; it prevents it from further commercialization and created a way for the finding of third-generation photon harvesters [6].

In this scenario, great efforts have been devoted on efficiently exploiting third-generation solar energy to meet environmental and energy demands. As one of the new generations of solar cells, inorganic semiconductors have thus recently been studied and attracted more attention in both industrial and academic fields. The outstanding characteristics of semiconductor quantum dots (QDs) in quantum-dot sensitized solar cells (QDSSCs) make them attractive due to their device stability and enhanced efficiency over DSSCs [7]. Moreover, the QDSSCs have tremendous advantages such as band gap tunability, high-absorption coefficient, multiple exciton generation possibilities and solution process ability. QDSSCs generally consist of a wide-band gap mesoporous oxide film, a redox electrolyte, QDs as a sensitizer and a counter electrode (CE) [8]. To enhance the light harvest in the visible light region, enough research has been conducted to synthesize the high-performance sensitizers. It has always been a challenge to prepare organic materials as effective sensitizers to absorb photons successfully under 1 sun illumination (AM 1.5G, $100\,mW/cm^2$). Due to this problem, narrow-band gap semiconductor QDs, such as PbS, CdSe, CdS and InAs have been used as a sensitizer on TiO_2 or ZnO as an alternative of organic dyes because of their versatile electrical and optical properties [9,10].

Even though QDSSCs have tremendous advantages, it still exhibits a moderate PCE of 6%–10% because of electron loss occurring at the CE/electrolyte and photoanode/electrolyte interface [11]. This lower PCE observed is independent of the configuration of QDSSCs, i.e., for both sensitized configuration and depleted heterojunction solar cells. Other probable factor for the limitation in PCE includes the difficulty of fabrication, enormous large numbers of semiconductor QDs on a n-type photo anode and the narrow absorption range of CdS QDs. Till date, a PCE of around 8.55% is achieved for the QD-depleted heterojunction solar cells with an active area of $1.37\,mm^2$ and solid-state based QDSSCs lie at 6.82% and 7.5% with an active area of 23.67 and $16\,mm^2$ [12]. Extensive research has been dedicated to increase the PCE of QDSSCs by two distinct approaches: on the one hand, improvement of short-circuit current density (J_{sc}) using enhanced light-harvesting capability or modification of sensitizers and on the other hand, enhancement of fill factor (FF) using CEs as exclusively adapted to polysulfide redox electrolyte, mainly CuS, CoS and NiS. However, improvement of open-circuit potential (V_{oc}) requires the enhancement of PCE in QDSSCs, which is possible by the use of new redox electrolyte or precise control of the recombination rate at interfaces [13,14].

In spite of the comparable V_{oc} and/or equal J_{sc} in QDSSCs, the reported *FF* for QDSSCs is still low compared with the DSSCs, which is normally more than 0.7. In recent years, great research has been carried out to improve *FF* and V_{oc} to enhance the *PCE* without sacrificing J_{sc}. Series resistance and charge recombination process at the CE/polysulfide electrolyte interface are the main factors that can noticeably affect the *FF* [15,16]. Suppressing recombination rate at the interface in DSSCs or QDSSCs is even more difficult task due to the inherent lower QD sensitizer loading on the TiO_2 or ZnO and potential contribution of defect states in QD surface. One of the main techniques to suppress charge recombination in sensitized cells is over coating the TiO_2 on the fluorine-doped tin oxide (FTO) substrate with a thin wide band gap inorganic barrier layer on the TiO_2 surface, which has been extensively explored in both DSSCs and QDSSCs. Our previous reports suggested that ZnS act as a most popular barrier later on the TiO_2 surface compared with bare TiO_2, Al_2O_3 and SiO_2 for preventing interfacial recombination in QDSSCs [17]. The barrier treatment on TiO_2 can be included before or after sensitization and all these inorganic barrier layers basically act as energy barriers to reduce the recombination process of photogenerated electrons in TiO_2 with holes residing in the hole conductor. Note that deposition of the barrier layer before or after, it provides a path for passivation of the QDs, affecting the electron transfer rate and enhancing the cell stability by protecting QDs from potential corrosion by the electrolyte. Moreover, an efficient electrochemical active material, high surface area, adequate surface roughness and narrow band gap semiconductor material for efficient CEs could achieve similar or even higher *FF* in QDSSCs [18].

The use of narrow band gap and p-type semiconductor material as a capable CE could lead to attain superior *FF* in QDSSCs. Platinum (Pt) proves to be a successful candidate as a CE for organic iodide/triiodide (I^-/I_3^-) redox electrolyte with appreciable catalytic activity and exhibiting minimum resistance [19]. However, the limited presence and high price of Pt prevent it from further commercialization in fuel cells and solar cell fields. Moreover, in conjunction with the aqueous polysulfide electrolyte, Pt and other metals like Au can be easily poisoned, leading to the rapid loss of the catalytic activity due to intermediate product which blocks the active sites of Pt in the polysulfide redox electrolyte, and sulfur containing compounds (S^{2-} or thiol) are strongly and preferentially adsorbed at the Pt surface, which is another real problem faced by most of the Pt-based electrocatalysts [20,21].

Therefore, great efforts have been made to investigate the Pt-free electrocatalysts to boost up the *PCE* of QDSSCs. There are many alternatives but compared to the precious metals, transition metals are very inexpensive, profuse, high electrocatalytic, exhibit lower resistance for the redox reaction and stress-free to be produced. Due to above advantages of transition metals, they are used as substitutes for noble metals in various applications with high effectiveness. The high surface area, roughness, structure, profile and particle size and composition of the transition metals have a straight effect on the catalytic performance and electrocatalytic applications in QDSSCs. The CE should have a low charge-transfer rate between the CE/electrolyte interface (R_{CE}), and it must play a major role in collecting the electrons from the external circuit and reducing the S_n^{2-} ions to S^{2-}. While there are many transition metals, copper, nickel, manganese and cobalt are the most preferred since they not only yield high *PCE* but also exhibit high proofed activity, cost-effective and superior photo and electrocatalytic properties and conductivity as well. The novel and efficient transition-metal sulfide-based composites strongly decrease the R_{CE} rate. Moreover, their optimized composition, time and temperature can distinctly improve the performance due to the adsorption capacity. New electronic structure and synergetic impact of CuS and MnS reveal good performance as CE electrode material in QDSSCs.

7.2 EXPERIMENTAL SECTION

7.2.1 PREPARATION OF CuS, MnS AND CuS/MnS COUNTER ELECTRODES

In this study, fluorine-doped tin oxide (FTO) was used as a substrate to deposit CuS/MnS active material due to its various advantages such as low-cost, high thermal stability, transparent, and conductive nature. The FTO substrates were cleaned with ethanol, acetone, and deionized water for 20 min each to remove some unwanted particles. The CuS, MnS and CuS/MnS electrode materials were coated on cleaned FTO substrates by using a single-step hydrothermal technique. In a typical experiment, 0.1 M $MnSO_4 \cdot H_2O$ and 0.1 M $CuSO_4$ are dissolved in 80 mL of deionized water and then stirred for 20 min. Later on, 0.4 M CH_4N_2O and 0.4 M $Na_2S_2O_3$ were added to the above solution and stirred for another 20 min to form a homogeneous solution. The thoroughly mixed solution was transferred to a Teflon-lined autoclave of 120 mL capacity and the cleaned FTO substrates were vertically placed and heated at 120°C for 6 h and then cooled naturally to room temperature. In order to prepare bare CuS and MnS electrodes for comparison with CuS/MnS composite, the above procedure was repeated in the absence of 0.1 M $MnSO_4 \cdot H_2O$ or 0.1 M $CuSO_4$ under the same experimental conditions.

7.2.2 PREPARATION OF TiO₂/CdS/CdSe/ZnS PHOTOANODES AND FABRICATION OF QDSSCs TO MEASURE J–V ANALYSIS

Titanium dioxide (TiO_2) were coated on cleaned FTO substrates using a commercially available 20 nm TiO_2 paste (Ti-nanoxide HT/SP, Solaronix) using the doctor-blade method with an active area of 0.25 cm². Subsequently, CdS, CdSe and ZnS passivation layers were also coated on top of TiO_2 using a successive ion layer adsorption and reaction method (SILAR). The detailed experimental procedure was explained in our previous studies [22]. Full cells of QDSSCs were prepared by sandwiching a counter electrode (CuS, MnS, and CuS/MnS) and photoanode (TiO_2/CdS/CdSe/ZnS) using a 60 µm sealant (SX 1170-60, Solaronix) at 90°C. The internal space between the counter electrode and photoanode was filled with a polysulfide electrolyte containing 2 M Na_2S, 2 M S and 0.2 M KCl and the ratio in methanol: water is 7:3.

7.2.3 CHARACTERIZATION

The crystalline structure of the prepared electrodes such as CuS, MnS and CuS/MnS were examined by X-ray diffraction (XRD, D/Max-2400, Rigaku) using a Cu K ∝ source and it is operated at 40 kV and 30 mA. X-ray photoelectron spectroscopy

(XPS, KBSI, Busan, South Korea) was carried out for the same electrodes to examine the chemical states of the elements using Al kα X-rays at 240 W. The surface morphology and elemental composition of the samples were evaluated using high-resolution scanning electron microscopy (HR-SEM, SU-70, Hitachi). The PCE was obtained using an ABET technologies (USA) solar simulator under 1 sun illumination (AM 1.5 G, 100 mW/cm²). Electrochemical impedance spectroscopy (EIS) and Tafel polarization analysis were performed on a symmetrical cell with a frequency ranging from 0.1 Hz to 500 kHz and a scan rate of 10 mV/S using a BioLogic potentiostat/galvanostat/EIS analyzer (SP-150, France).

7.3 RESULTS AND DISCUSSION

The surface morphology of the prepared electrode materials such as CuS, MnS and CuS/MnS counter electrodes were studied using HR-SEM. Figure 7.1 shows the top surface HR-SEM images of the CuS, MnS and CuS/MnS counter electrodes (CEs). The HR-SEM images show various surface structures with different metal sulfide materials and their composite CEs. Figure 7.1a and b shows the HR-SEM images of CuS and MnS on nickel foam (NF) substrates, exhibits a coral reef and clustered coral reef like nanostructure and this clustered were tightly connected to each other. However, these coral reef and clustered coral reef like nanostructures of CuS and MnS were peel off from the substrate and it is very vital for the active material to strongly adhere on the substrate to enhance the PCE of the device. Figure 7.1c shows the HR-SEM image of CuS/MnS CE exhibits a cone-shaped nanostructure without any void places on nickel foam substrates and shows greater adhesion. This kind of nanostructure can enable superior surface area and enriched surface morphology will increase efficient electron transfer across the counter electrode and polysulfide redox electrolyte [23,24].

Energy-dispersive X-ray spectroscopy (EDX) analysis finds the elemental composition of bare CuS and MnS electrodes on nickel foam substrates. As shown in Figure 7.2a, EDX conforms the presence of copper (Cu), sulfide (S), nickel (Ni) with larger peak intensity and also presence the oxygen (C) and carbon (C) with less intensity. Figure 7.2b affirms the presence of Ni, S, C, O and Manganese (Mn). The atomic percentage of Cu, Ni, S, O and C was approximately 15.55%, 24.39%, 28.8%, 11.39% and 19.88% for CuS-based films and 0.56% (Mn), 40.16% (Ni), 29.58% (S), 13.45% (O) and 16.25% (C) for MnS CEs. This EDX result illustrates the successful deposition of active elements on nickel foam substrates.

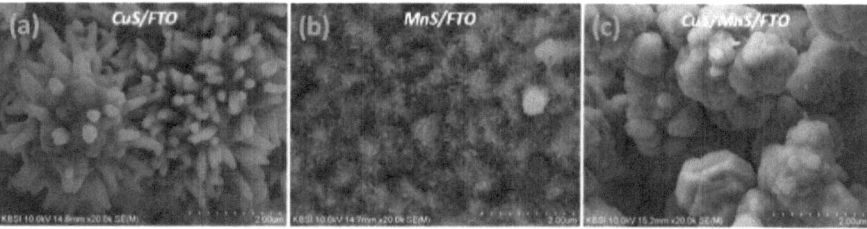

FIGURE 7.1 High-resolution scanning electron microscopy (HR-SEM) images of (a) CuS, (b) MnS and (c) CuS/MnS on nickel foam substrates.

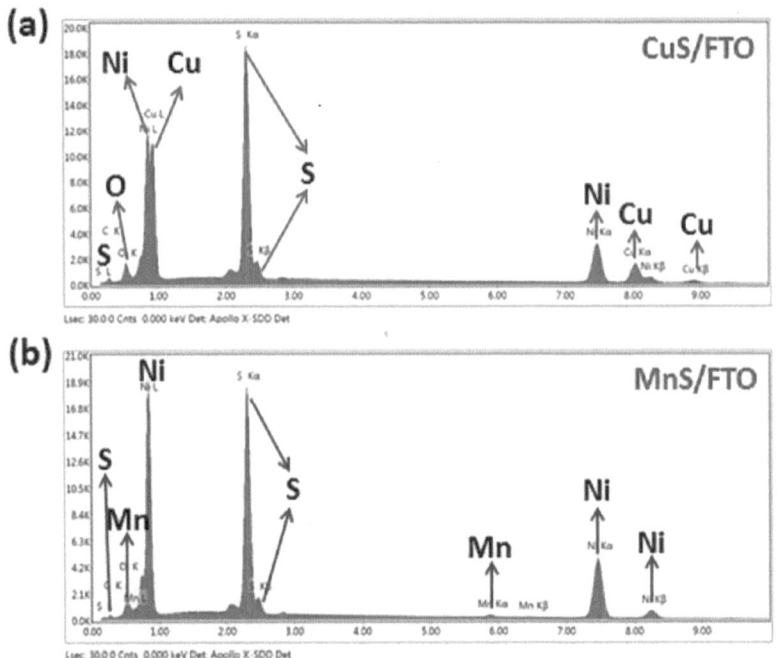

FIGURE 7.2 Energy-dispersive X-ray spectroscopy spectrum of the bare (a) CuS and (b) MnS materials on nickel foam substrates.

XRD is a rapid analytical technique and used for phase identification of a crystal-line material. Figure 7.3 shows the XRD spectrum of the CuS/MnS material on the nickel foam substrate. The three strong peaks observed at 44.5°, 51.8° and 76.4° for the nickel foam substrate. The peaks observed at 21.78°, 31.82°, 43.14° and 53.27° for

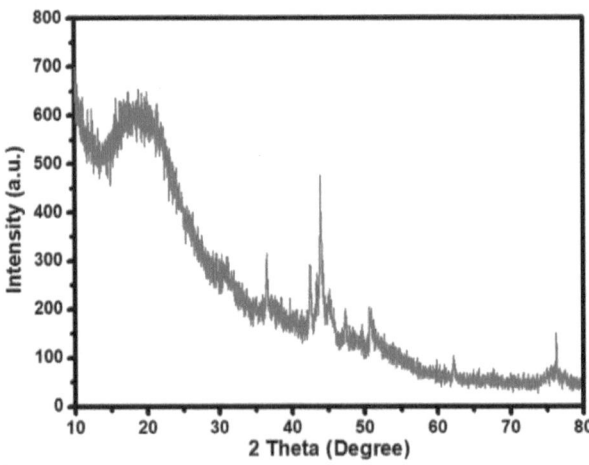

FIGURE 7.3 X-ray diffraction spectrum of CuS/MnS on nickel foam substrates.

the CuS and 28.53°, 34.30°, 42.61°, 56.43° and 60.8° for the MnS and these obtained peaks were well matched with the JCPDS No: 01-079-2321, 40-1288 and 40-1289. Therefore, the EDX and XRD results suggested that the CuS and MnS materials had been deposited successfully onto the FTO substrate using the facile hydrothermal method.

To further characterize the electrodes, the ultraviolet–visible (UV–vis) absorption spectra of different counter electrodes were investigated and the curves are shown in Figure 7.4. When CuS/MnS was coated on the FTO substrate using the hydrothermal method, it extends the absorption spectrum to the wavelength exceeding the infra-red region compared to bare CuS and MnS thin films. Furthermore, the absorbance slightly increased with the combination of bimetallic sulfide materials, indicating that the Cu and Mn particles show positive interactions and can increase the thickness of the composite thin film. The UV–vis spectrum of the CuS/MnS/FTO film never reaches zero intensity but extends the absorption spectrum to the wavelength exceeding 1,000 nm, which is due to free-carrier intraband absorption. Moreover, the CuS/MnS/FTO composite shows strong absorption intensity and wavelength values and lies beyond the instrumental limitation. The asset of absorbing higher wave-length irradiation generates CuS/MnS/FTO a convenient counter electrode material for harvesting the residual light and a positive shift of the Quasi-Fermi level and enhances the open-circuit voltage of QDSSCs.

For obtaining higher photovoltaic properties of the prepared QDSSCs, electro-chemical properties such as series resistance, interfacial charge transfer, charge recombination between counter electrode/electrolyte and photoanode/electrolyte, and Warburg diffusion resistance in the polysulfide redox electrolyte are very impor-tant. EIS was performed for two identical CEs separated by a polysulfide redox electrolyte under the dark condition. Figure 7.5 shows the corresponding Nyquist

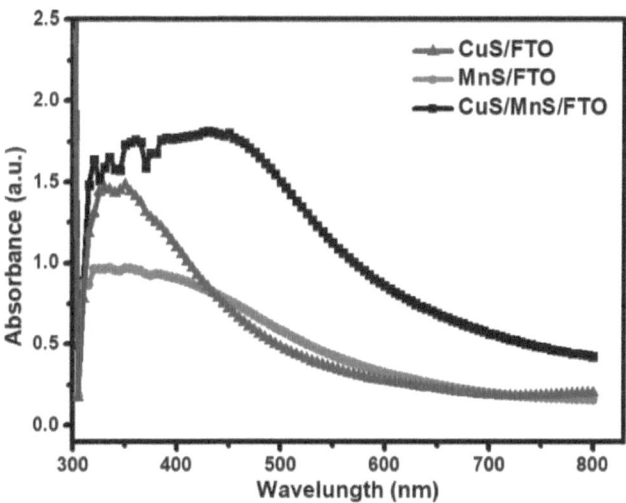

FIGURE 7.4 UV absorption spectra of CuS, MnS and CuS/MnS thin films deposited on fluorine-doped tin oxide (FTO) substrates.

plots of CuS, MnS and CuS/MnS/FTO composite materials and it is recorded at an open-circuit potential in the frequency range from 10^{-1} to 5×10^5 Hz. As shown in Figure 7.5, all QDSSCs consist of two semicircles and it provides the information about the charge-transfer resistance (R_{ct}) at the counter electrode/polysulfide redox electrolyte interface and Warburg diffusion resistance. The CuS/MnS/FTO composite CE shows smaller series resistance ($R_s = 7.64$ Ω) value compared to other two electrodes (8.31 Ω for CuS/FTO and 9.84 Ω for MnS/FTO), indicating that the CuS/MnS active material strongly adhered to the FTO and help the collection of electrons from the external circuit. The R_{ct} values calculated for the bare CuS and MnS/FTO CEs are much higher than that of the CuS/MnS/FTO composite. The much lower R_{ct} values for the CuS/MnS/FTO composite demonstrate a superior electrocatalytic activity toward the polysulfide redox electrolyte and higher charge-transfer process at the counter electrode (CuS/MnS)/polysulfide redox electrolyte interface [25]. The Warburg diffusion value also shows the same trend for series resistance and charge-transfer resistance between counter electrode/electrolyte with a much lower value for the CuS/MnS/FTO composite than that of bare CuS and MnS/FTO [26,27]. As per previous reports, the lower Z_w value may be obtained due to the smooth surface morphology [as seen in the HR-SEM image (Figure 7.1)], which guarantees complete use of the CuS/MnS/FTO composite for the reduction of oxidized polysulfide.

Figure 7.6 shows the J–V curves of QDSSCs assembled with CuS, MnS and CuS/MnS counter electrodes and TiO$_2$/CdS/CdSe/ZnS photoanode at 100 mW/cm^2 light intensity and the corresponding photovoltaic parameters such as *PCE*, *FF*, open-circuit voltage (V_{oc}) and short-circuit current density (J_{sc}) are listed in Table 7.1. The *PCE* and *FF* of QDSSCs are determined by using Equations 7.1 and 7.2.

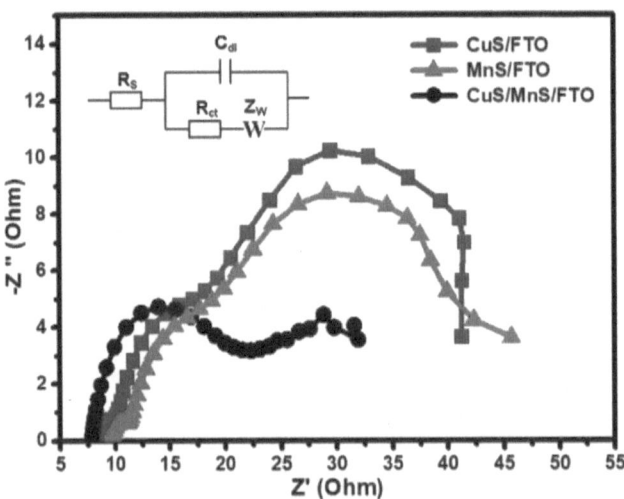

FIGURE 7.5 Nyquist plots for the symmetric cell of two identical counter electrodes of CuS, MnS and CuS/MnS/FTO composites. The inset shows the equivalent circuit to fit Nyquist plots.

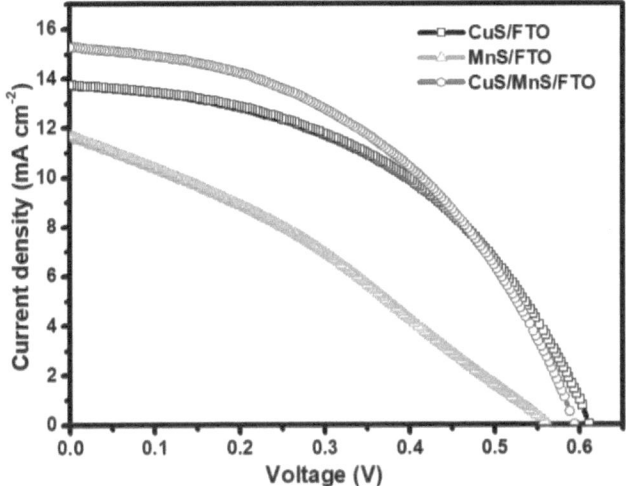

FIGURE 7.6 Comparison of photocurrent density voltage (J–V) curves of QDSSCs assembled with CuS, MnS and CuS/MnS counter electrodes under 1 sun illumination (100 mW/cm²).

TABLE 7.1

Photovoltaic Performance of the CuS, MnS and CuS/MnS/FTO Composite CE-Based QDSSCs

CE Material	V_{oc} (V)	J_{sc} (mA/cm²)	FF	PCE (%)
CuS/FTO	0.61	13.29	0.46	3.71
MnS/FTO	0.56	12.32	0.31	2.02
CuS/MnS/FTO	0.60	15.55	0.46	4.38

$$P_m = J_m \times V_m \tag{7.1}$$

$$FF = \frac{P_m}{J_{SC} \times V_{OC}} = \frac{J_m \times V_m}{J_{SC} \times V_{OC}} \tag{7.2}$$

$$PCE(\%) = \frac{P_m}{P_{in}} = \frac{FF \times J_{SC} \times V_{OC}}{P_{in}} \tag{7.3}$$

where J_m is the maximum product of current density, V_m is the maximum voltage, J_{sc} is the short-circuit current density, V_{OC} is the open-circuit voltage, I_{SC} is the short-circuit current, and P_{in} is the input power. As shown in Figure 7.6 and Table 7.1, the photovoltaic performance of the QDSSCs fabricated using CuS/MnS/FTO composite shows higher compared to bare CuS and MnS counter electrodes. The photovoltaic performance of QDSSCs with CuS/FTO is superior to that of the MnS/FTO with V_{oc}, J_{sc}, FF and PCE values of 0.56 V, 12.32 mA/cm², 0.31% and 2.02% for

MnS and 0.61 V, 13.29 mA/cm², 0.46 and has the highest *PCE* of 3.71% for CuS/FTO CEs. On the other hand, the *PCE* (4.38%) and J_{sc} (15.55 mA/cm²) of the CuS/MnS/FTO composite CEs were much higher than that of other two electrodes, which were obtained to the superior electrocatalytic ability of CuS/MnS/FTO composite in the reduction of polysulfide redox electrolytes used in QDSSCs [28]. The combination of two different metallic sulfide materials can improve the electrical conductivity considerably because of their positive synergetic effect and reduced recombination rate. The higher photovoltaic performance of CuS/MnS/FTO composite is due to various reasons such as reduced recombination rate, charge-transfer resistance at the counter electrode/polysulfide redox electrolyte and also reduced internal resistance [29].

According to the EIS model, the electron lifetime can be calculated from the angular frequency of the impedance semicircle arc at the mid-frequency (7.4):

$$\tau_e = \frac{1}{2\pi f_{max}} \tag{7.4}$$

Here, τ_e is the electron lifetime, and f_{max} is the maximum frequency of the bode-plot curves. The corresponding curves are shown in Figure 7.7. As shown in Figure 7.7, the τ_e value increases from 1.87 to 2.26 ms as the MnS particles incorporated with CuS onto FTO substrates. The improved electron lifetime is due to the lower recombination rate of the injected electrons for the reduction of the electrolyte, synergetic effect and increased electrical conductivity with the composite-based electrode [30,31]. When QDSSCs fabricated with bare CuS- or MnS-based electrode materials, the electron lifetime decreases visibly. This may be because of higher diffusion coefficient value and higher recombination rate. According to the results of the EIS and bode-plot (electron transfer rate and lifetime), the appropriate composite materials

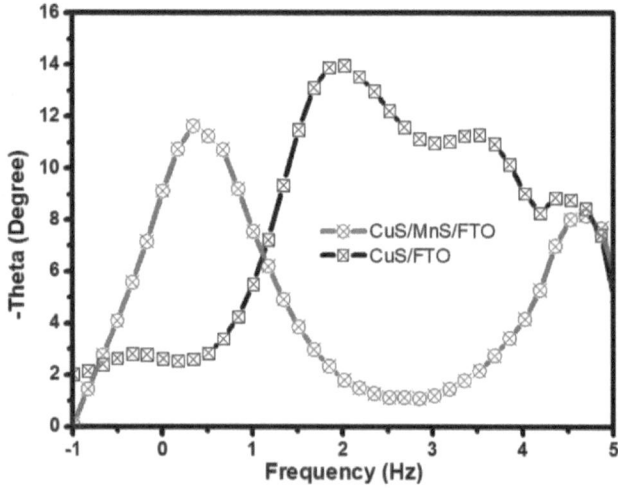

FIGURE 7.7 Frequency versus phase (Bode-plot) of the QDSSCs of the CuS and CuS/MnS/FTO counter electrodes.

and/or their concentration can drastically affect their electrochemical properties and the obtained results are consistent with the results of photovoltaic parameters.

7.4 CONCLUSIONS

CuS, MnS and CuS/MnS composite materials were successfully deposited on FTO substrates using the facile hydrothermal method. The as-obtained CuS/MnS/FTO composite was used as an efficient counter electrode material in QDSSCs along with a TiO_2/CdS/CdSe/ZnS as the photo anode and polysulfide redox as the electrolyte. The effect of the two different metal sulfide materials and their combination as a composite on the performance of the QDSSCs was studied using various analyses such as HR-SEM, XRD, EIS, J–V and bode-plot. The J–V results of the QDSSCs were studied and the results concluded that the CuS/MnS/FTO composite showed a higher PCE of 4.38% than that using bare CuS (3.71%) and MnS (2.02%), under 1 sun illumination (100 mW/cm^2, AM 1.5 G). EIS studies also exhibited a lower charge-transfer resistance between the CE/electrolyte for the CuS/MnS/FTO composite than the other two materials. The obtained $V_{oc} = 0.60$ V, $J_{sc} = 15.55$ mA/cm^2, $FF = 0.46$ and *PCE* of 4.38% of the CuS/MnS/FTO composite counter electrode resulted from the nice and uniform surface morphology on FTO which helps in faster electron transport with a lower charge recombination rate. These obtained results suggest that the CuS/MnS/FTO composite can be an alternative efficient counter electrode material to Pt CEs.

ACKNOWLEDGMENT

This research was supported by the National Research Foundation of Korea (NRF) grant funded by the Korea government (Ministry of Science and ICT) (2020R1G1A1010247).

REFERENCES

1. M.A. Green, Third generation photovoltaics: Ultra-high conversion efficiency at low cost, *Prog. Photovolt. Res. Appl.* 9 (2001) 123.
2. P.V. Kamat, Quantum dot solar cells. Semiconductor nanocrystals as light harvesters, *J. Phys. Chem. C* 112 (2008) 18737.
3. C.D. Sunesh, C.V.V. Gopi, M.P.A. Muthalif, H.J. Kim and Y. Choe, Improving the efficiency of quantum-dot-sensitized solar cells by optimizing the growth time of the CuS counter electrode, *Appl. Surf. Sci.* 416 (2017) 446–453.
4. Z. Du, M. Artemyev, J. Wang and J. Tang, Performance improvement strategies for quantum dot-sensitized solar cells: A review, *J. Mater. Chem. A* 7 (2019) 2464–2489.
5. Y. Wu, D. Yan, J. Peng, T. Duong, Y. Wan, P. Phang, H. Shen, N. Wu, C. Barugkin, X. Fu, S. Surve, D. Walter, T. White, K. Catchpole and K. Weber, Monolithic perovskite/silicon-homojunction tandem solar cell with over 22% efficiency, *Ener. Environ. Sci.* 10 (2017) 2472–2479.
6. A.E. Reddy, S.S. Rao, C.V.V.M. Gopi, T. Anitha, C.V.T. Varma, D. Punnoose and H.J. Kim, Morphology controllable time-dependent CoS nanoparticle thin films as efficient counter electrode for quantum dot-sensitized solar cells, *Chem. Phys. Lett.* 687 (2017) 238–243.

7. H.J. Kim, C.W. Kim, D. Punnoose, C.V.V.M. Gopi, S.K. Kim, K. Prabakar and S.S. Rao, Nickel doped cobalt sulfide as a high performance counter electrode for dye-sensitized solar cells, *Appl. Surf. Sci.*, 328 (2015) 78–85.

8. B. O'Regan and M. Gratzel, A low-cost, high efficiency solar cell based on dye sensitized solar cells colloidal TiO_2 films, *Nature* 353 (1991) 737–740.

9. W.S. Chi, J.K. Koh, S.H. Ahn, J.S. Shin, H. Ahn, D.Y. Ryu and J.H. Kim, Highly efficient I_2-free solid-state dye-sensitized solar cells fabricated with polymerized ionic liquid and graft copolymer-directed mesoporous film, *Electrochem. Commun.* 13 (2011) 1349–1352.

10. J-J. Lagref, Md.K. Nazeeruddin and M. Gratzel, Molecular engineering on semiconductor surfaces: Design, synthesis and application of new efficient amphiphilic ruthenium photosensitizers for nanocrystalline TiO_2 solar cells, *Synthet. Metals* 138 (2003) 333–339.

11. H.J. Kim, J.H. Kim, CH.S.S.P. Kumar, D. Punnoose, S.K. Kim, C.V.V.M. Gopi and S.S. Rao, Facile chemical bath deposition of CuS nano peas like structure as a high efficient counter electrode for quantum-dot sensitized solar cells, *J. Electroanalyt. Chem.* 739 (2015) 20–27.

12. D. Punnoose, H.J. Kim, S.S. Rao and CH.S.S.P. Kumar, Cobalt sulfide counter electrode using hydrothermal method for quantum dot-sensitized solar cells, *J. Electroanalyt. Chem.* 750 (2015) 19–26.

13. J. Liu, C. Wang, Z. Ge, T. Wang, L. Xia, Y. Wu, et al., Investigation of the CdS quantum dot sensitized solar cells based on a series of zinc titanium mixed metal oxides, *Opt. Mater.* 107 (2020) 110059.

14. Y. Bai, C. Han, X. Chen, H. Yu, X. Zong, Z. Li and L. Wang, Boosting the efficiency of quantum dot sensitized solar cells up to 7.11% through simultaneous engineering of photocathode and photoanode, *Nanomater. Ener.* 13 (2015) 609–619.

15. A. Sahasrabudhe and S. Bhattacharyya, Dual sensitization strategy for high-performance core/shell/quasi-shell quantum dot solar cells, *Chem. Mater.* 27 (2015) 4848–4859.

16. H.-J. Kim, T.-B. Yeo, S.-K. Kim, S.S. Rao, A.D. Savariraj, K. Prabakar and C.V.V. M. Gopi, Optimal-temperature-based highly efficient NiS counter electrode for quantum-dot-sensitized solar cells, *Eur. J. Inorg. Chem.* (2014) 4281–4286.

17. S.S. Rao, I.K. Durga, C.V.V.M. Gopi, C.V. Tulasivarma, S.K. Kim and H.J. Kim, The effect of TiO_2 nanoflowers as a compact layer for CdS quantum-dot sensitized solar cells with improved performance, *Dalton Trans.* 44 (2015) 12852–12862.

18. L. Wen, T. Huang, M. Huang, Z. Lu, Q. Chen, Y. Meng and L. Zhou, Secondary reduction of graphene improves the photoelectric properties of TiO_2@rGO composites, *Ceram. Int.* 46 (2020) 9930–9935.

19. C.V. Thulasi-Varma, S.S. Rao, CH.S.P. Kumar, C.V.V.M. Gopi, I.K. Durga, S.K. Kim, D. Punnoose and H.J. Kim, Enhanced photovoltaic performance and time varied controllable growth of a CuS nanoplatelet structured thin film and its application as an efficient counter electrode for quantum dot-sensitized solar cells via a cost-effective chemical bath deposition, *Dalton Trans.* 44 (2015) 19330–19343.

20. I. Robel, V. Subramanian, M. Kuno and P.V. Kamat, Quantum dot solar cells Harvesting light energy with CdSe nanocrystals molecularly linked to mesoscopic TiO_2 films, *J. Am. Chem. Soc.* 128 (2006) 2385.

21. M. Sabet, M.S. Niasari and O. Amiri, Using different chemical methods for deposition of CdS on TiO_2 surface and investigation of their influences on the dye-sensitized solar cell, *Electrochim. Acta* 117 (2014) 504.

22. I.K. Durga, S.S. Rao, D. Punnoose, N. Kundakarla, C.V. Tulasivarma and H.J. Kim, An innovative catalyst design as an efficient electro catalyst and its applications in quantum-dot sensitized solar cells and the oxygen reduction reaction for fuel cells, *New J. Chem.*, 41 (2017) 2098–2111.

23. S.S. Rao, I.K. Durga, N. Kundakarla, D. Punnoose, C.V.V.M. Gopi, A.E. Reddy, M. Jagadeesh and H.J. Kim, A hydrothermal reaction combined with a post anion-exchange reaction of hierarchically nanostructured NiCo$_2$S$_4$ for high-performance QDSSCs and supercapacitors, *New J. Chem.*, 41 (2017) 10037–10047.

24. H.M. Choi, I.A. Ji and J.H. Bang, Metal selenides as a new class of electrocatalysts for quantum dot sensitized solar cells: A tale of Cu 1.8 Se and PbSe, *ACS Appl. Mater. Interfaces* 6 (2014) 2335–2343.

25. J. He, D. Wu, Z. Gao, F. Xu, S. Jiang, S. Zhang, K. Cao, Y. Guo and K. Jiang, Graphene sheets anchored with high density TiO$_2$ nanocrystals and their application in quantum dot-sensitized solar cells, *RSC Adv.* 4 (2014) 2068–2072.

26. H.J. Kim, J.H. Kim, I.K. Durga, D. Punnoose, N. Kundakarla, A.E. Reddy and S.S. Rao, Densely packed zinc sulfide nanoparticles on TiO$_2$ for hindering electron recombination in dye-sensitized solar cells, *New J. Chem.* (2016) (40) 9176–9186.

27. L. Wen, Q. Chen, L. Zhou, M. Huang, J. Lu, X. Zhong, H. Fan and Y. Ou, Cu1.8S@CuS composite material improved the photoelectric efficiency of cadmium-series QDSSCs, *Opt. Mater.* 108 (2020) 110172.

28. X. Huang, S. Huang, Q. Zhang, X. Guo, D. Li, Y. Luo, Q. Shen, T. Toyoda and Q. Meng, A flexible photoelectrode for CdS/CdSe quantum dot-sensitized solar cells(QDSSCs), *Chem. Commun.* 47 (2011) 2664.

29. J. Wu, Q. Li, L. fan, Z. Lan, P. Li, J. Lin and S. Hao, High-performance polypyrrole nanoparticles counter electrode for dye-sensitized solar cells, *J. Power Source.* 181 (2008) 172–176.

30. M. Gratzel, Photoelectrochemical cells, *Nature* 414 (2001) 338–344.

31. T. Renourad, R.-A. Fallahpour, Md.K. Nazeeruddin, H.R. Baker and M. Gratzel, Novel ruthenium sensitizers containing functionalized hybrid tetradentate ligands: Synthesis, characterization, and INDO/S analysis, *Inorgan. Chem.* 41 (2002) 367–378.

8 Performance Investigation of LG-SDM-FSO Transmission Link under Snow Weather Conditions

Mehtab Singh
Satyam Institute of Engineering and Technology

Meet Kumari
Chandigarh University

Anu Sheetal
Guru Nanak Dev University

Reecha Sharma
Punjabi University

Jyoteesh Malhotra
Guru Nanak Dev University

Amit Grover
Shaheed Bhagat Singh State University

CONTENTS

8.1   INTRODUCTION

Wireless communication systems include independent radio base stations along with network operators that result in an extravagant infrastructure, unworkable equipment and ineffective cellular architecture. In 2015, the increasing number of mobile customers touched 7.2 billion, as stated by International Telecommunication Union (ITU). By 2021, a sharp rise of approximately 65% in annual mobile data traffic is seen which could be enhanced ten times [1,2]. Besides, this resulted in wireless operators restricting the radio frequency (RF) spectrum allocation. Also, the growth in mobile consumers opened possibilities for radio over free space optics (Ro-FSO) which transmits the multiple RF signals over a high-data rate optical carrier. Ro-FSO does not desire excessive investment in the optical fiber cable and overcomes the RF signals' congestion issue which is another serious concern about latest wireless communication systems [3,4].

Ro-FSO incorporates the radio-over-fiber (RoF) technology and free space optics (FSO) technologies. It characterizes high-speed transmission with less attenuation and power consumption which can again be used to focus several processes like up–down conversion, switching, RF hand-off, coding and multiplexing [5,6]. FSO link is a wireless optical technology that utilizes light propagation in the free space to send information without cables/wires for the internet, computer networking and telecommunications. FSO connection is prepared to create a high-security, high-speed and friendly connection with a greater bandwidth light signal. It has several attractive features such as easy installation, flexible and practical links because of their inherent resistance to electromagnetic interference and license-free function [7,8]. As the favorability of broadband internet, computer networking and telecommunications services etc. rises, so the does the data centers transmission loading of data. As a result of the widespread use of 4G/5G mobile communications, customers are growing, and there is large transmission loading in information centers [9,10].

Furthermore, Ro-FSO can utilize economic, high-speed connection and mobile connectivity to integrate wireless network, RF and recent wireless communication system. But various atmospheric conditions consisting of rain, fog etc. can impact the Ro-FSO system. Mode division multiplexing (MDM) is a significant technology that can improve the optical system transmission capacity and used by researchers in different communication systems [3]. The several modes used in MDM are donut, Laguerre-Gaussian (LG) and most efficient Hermite-Gaussian (HG) modes [11–13]. The exponential rise of information traffic in Ro-FSO has encouraged researchers worldwide to explore several multiplexing schemes within optical access networks to enhance its transmission distance and capacity.

Mostly multiplexing techniques for access networks are based upon time division multiplexing (TDM), wavelength division multiplexing (WDM) and optical code division multiplexing (OCDM). But space division multiplexing (SDM) is a novel scheme that has fascinated remarkable notice as an advanced technology which enhances the transmission rate [14]. In SDM, spatial modes are used as data channels transferring independent information streams. Moreover, it is a rapidly emerging as a capable candidate for improving the total bandwidth of latest optical communication systems. SDM consists of MDM and the key idea behind SDM is the propagation of multiple spatial modes through the FSO channel. Channel capacity can be manifold by the number of spatial modes [15]. Although SDM-based FSO has various merits, its performance is disastrously affected by the weather as well as atmospheric turbulences. The major reasons of interferences are shown which affects the features of the transmitted light and decreases the transmitted signal power density and FSO range [16].

To fulfill the demand, Sara Alshwani et al. investigated the ten channels SDM over 4.2 and 1.1 km FSO link at 100 Gbps transmission rate under the influence of heavy rain and heavy haze, respectively [16]. Mohammed Nasih Ismael et al. reported the 12-channel SDM over FSO link for fiber to the home applications FTTH applications [15]. Sushank Chaudhary and Angela Amphawan demonstrated the MDM Ro-FSO system employing LG00, LG01 and LG02 modes in association with solid core photonics crystal fibers (SC-PCFs) over 2.5 km Ro-FSO link carrying 2.5 Gbps–5 GHz data under the impact of fog [3]. Aras Al-Dawoodi et al. investigated the eight-channel 40 Gbps MDM-FSO system over 3,000 and 650 m FSO link under clear and heavy rain atmospheric conditions [17]. Mehtab Singh et al. proposed and evaluated the 100 Gbps polarization division multiplexing (PDM)/CO-orthogonal frequency division multiplexing (OFDM) FSO system over 15 km FSO link under the impact of adverse climate conditions [18]. Vigneswaran Dhasarathan et al. reported the 16-level quadrature amplitude modulation (PDM-16-QAM)-based FSO link at 160 Gbit/s under dynamic weather conditions [19]. Heyam Maraha et al. investigated the 110 Gbps dense wavelength division multiplexing (DWDM)-FSO system under different weather conditions like clear air, haze, and heavy haze for 9.5, 3 and 2.5 km FSO distances, respectively [5]. Maged Abdullah Esmail et al. investigated and demonstrated the 1.08 Tbps hybrid FSO/fiber system over 100 m FSO link under the impact of light dust storm condition, background noise and misalignment effect [20]. Mohammed Nasih Ismae et al. reported the 140 Gbps hybrid coarse wavelength division multiplexing (CWDM) and DWDM system over 2.6 and 1.6 km FSO link under moderate and heavy rain weather conditions, respectively [21]. Angela Amphawan et al. reported 5 Gbps MDM-Ro-FSO system employing HG01 and HG03 modes over 8 km FSO link [22]. Kai Pang et al. experimentally demonstrated 400 Gbps quadrature-phase-shift keyed (QPSK) FSO system employing HG01, HG10, HG03, HG30, LG10, LG10, LG11 and LG-11 modes in a lateral misalignment [23]. M. Balasaraswathi et al. reported the successful transmission of 20 Gbps–40 GHz WDM MDM Ro-FSO system over 2,000 m FSO link using four beams over heavy fog condition incorporating HG and LG modes [24].

Also, the amount of information that is transmitted continues increasing. To handle this, several methods are used such as modulation techniques, appropriate light source, transmitting power levels estimation and transmission of wavelengths [15].

This chapter aims at utilizing system which has not been proposed and analyzed, to the best of our knowledge, in either one the pervious works. It utilizes 10 Gbps–10 GHz SDM-based Ro-FSO system employing LG00 and LG01 modes under dry and wet snow weather conditions. This chapter is organized as: Section 8.2 presents the SDM-based Ro-FSO transmission link using which two independent 10 Gbps–10 GHz data streams are transmitted concurrently over two LG00 and LG01 modes of a single laser beam. Section 8.3 describes the results and discussions of the system. Finally, conclusion is drawn in Section 8.4.

8.2 SIMULATION SETUP

In this chapter work, Optisystem v.15 simulation software is utilized to design and evaluate the LG-SDM-FSO transmission link under snow weather conditions incorporating two distinct LG spatial modes viz. LG00 and LG01 over a FSO link. Figure 8.1 demonstrates the diagram of the proposed Ro-FSO system.

Two distinct NRZ line encoded signals each consisting of 10 Gbps data are contemporaneously transferred over LG00 and LG01 modes at an operating wavelength of 1,550 nm which can be expressed (Figure 8.2) [14]:

$$\emptyset_{a,b}(x,y)$$

$$= \left(2r^2 \Big/ \omega_{0,x}^2 \right) L_a^b exp\left(\frac{2x^2}{\omega_{0x}^2} \right) exp\left(j\frac{r^2}{\omega_{0x}^2} \right) exp\left(j\frac{\pi x^2}{\lambda R_{0y}} \right) \left\{ \begin{array}{l} sin(ny), n > 0 \\ cos(ny), \ n > 0 \end{array} \right. \tag{8.1}$$

where x and y denote the mode dependencies on X- and Y-polarization axis, respectively, R is the curvature radius, ω represents the spot size, and L is the Laguerre polynomial.

For generating two independent LG00 and LG01 modes at the central wavelength of 1,550 nm, two spatial continuous wave (CW) lasers in addition with multi-mode generators (MMGs) are utilized. The generated 10 Gbps binary information per spatial laser utilizing a Pseudo-random bit sequence generator (PRBS) is line encoded through a NRZ encoder and further forwarded toward a Mach-Zehnder modulator (MZM) [25–27]. After the MZM modulation of the incoming signals over an optical carrier signal of 1,550 nm wavelength, both output MZM signals are combined using an MDM multiplexer (MUX) and transmitted over the FSO channel incorporating transmitting lens. Likewise, two independent LG00 and LG01 modes are obtained and passed through WDM MUX [28–30]. The spatial profiles of the transmitted signal of LG00, LG01 and LG00+LG01 modes are depicted in Figure 8.3a–c, respectively.

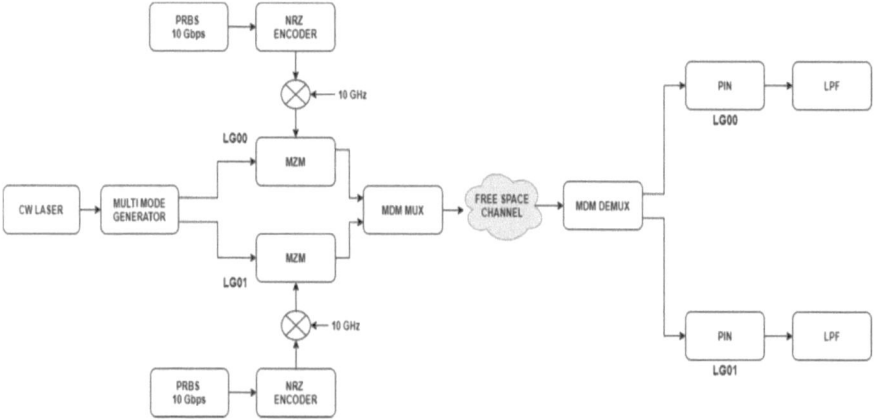

FIGURE 8.1 Simulation setup.

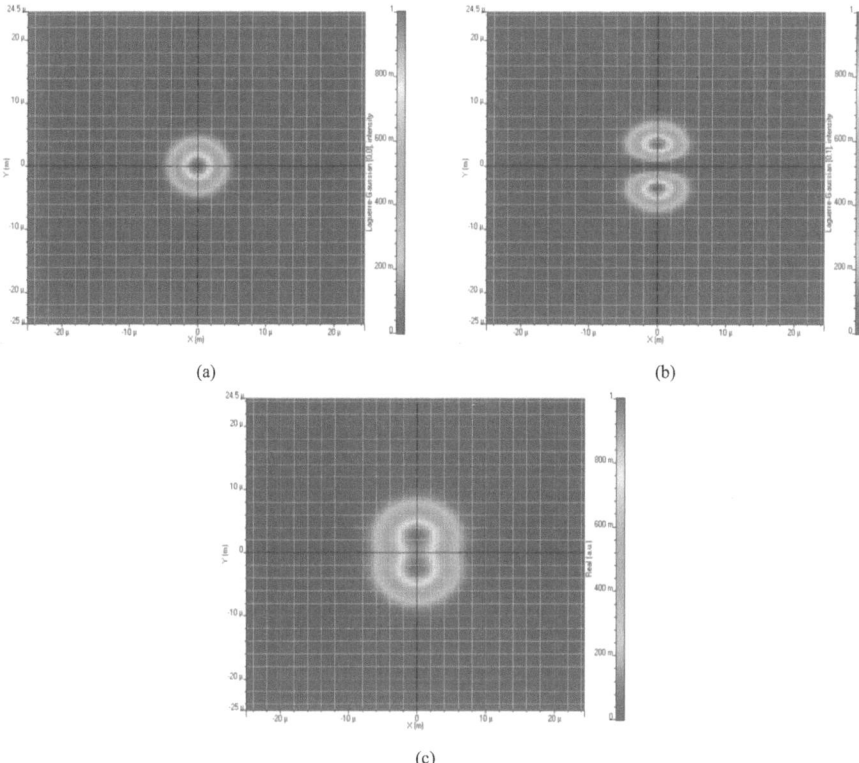

FIGURE 8.2   (a) LG00 mode, (b) LG01 mode and (c) LG00+LG01 mode.

On the receiver side, distinct LG00 and LG01 modes are obtained using a WDM demultiplexer (DEMUX) followed by two MDM DEMUX. A PIN photodiode is used to recover the received signal, low pass filter (LPF) to reduce the noise and to attain the high-quality signal, bit-error rate (BER) analyzer are used. The signal quality is analyzed through a BER analyzer in terms of BER and eye diagrams. Table 8.1 presents the simulation parameters used in this work.

8.3 RESULTS AND DISCUSSION

Figure 8.4 shows the performance of the Ro-FSO link-based LG-SDM-FSO transmission link system for NRZ modulation wavelengths considering transmitting and receiving aperture diameter 5 and 20 cm, respectively, and 1.5 urad divergence angle in terms of log of BER as performance metrics under the impact of wet and dry snow.

From the obtained results in Figure 8.3, it can be depicted that the log of BER of the received signal for LG00 mode channel is computed as −52, −29 and −16 over the Ro-FSO range of 700, 750 and 800 m, respectively, at an input power of 10 dBm. Also, for LG01 mode channel, the log of BER of the received signal is determined as −25, −14 and −8 over the FSO range of 700, 750 and 800 m, respectively. Similarly, as shown in Figure 8.3, at the same Ro-FSO range, for LG00 and LG01

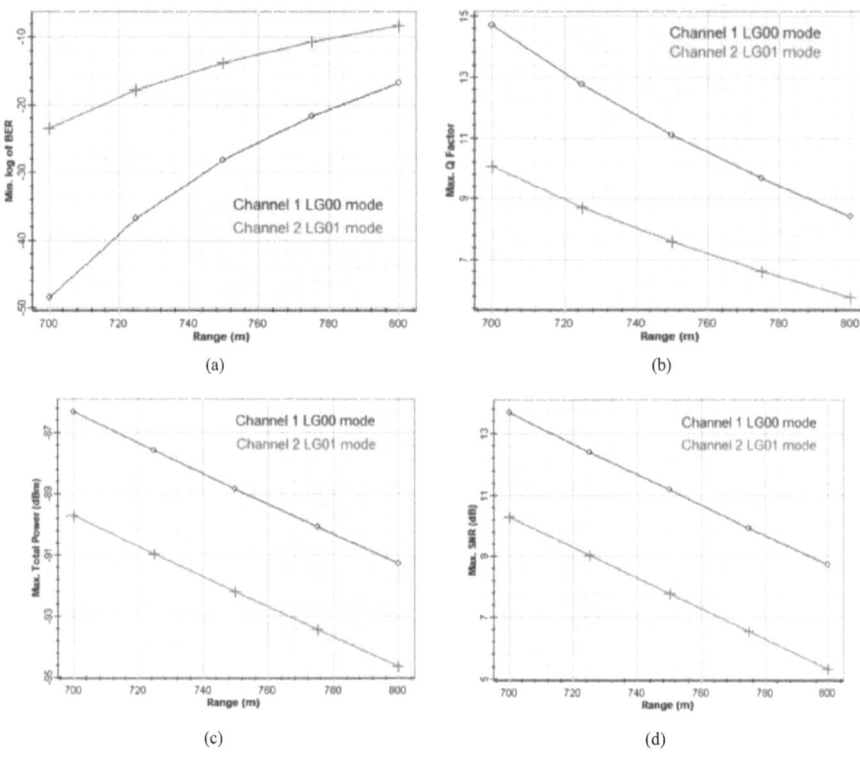

(a)

(b)

(c)

(d)

FIGURE 8.3 (a) Log of BER, (b) Q factor, (c) power, and (d) SNR versus range under wet snow.

TABLE 8.1
Simulation Parameters of LG-SDM-FSO Transmission Link

Parameter	Value
Laser frequency (THz)	193.1
Laser power (dBm)	10
Laser linewidth (MHz)	0.1
Bit rate/channel (Gbps)	10
Transmitter aperture diameter (cm)	5
Receiver aperture diameter (cm)	20
Divergence angle (mrad)	1.5
PIN responsivity (A/W)	1
Dark current (nA)	10
Sequence length	1,024
Samples per bit	32
Attenuation coefficients (dB/km)	13.73 (for wet snow)
	96.8 (for dry snow)

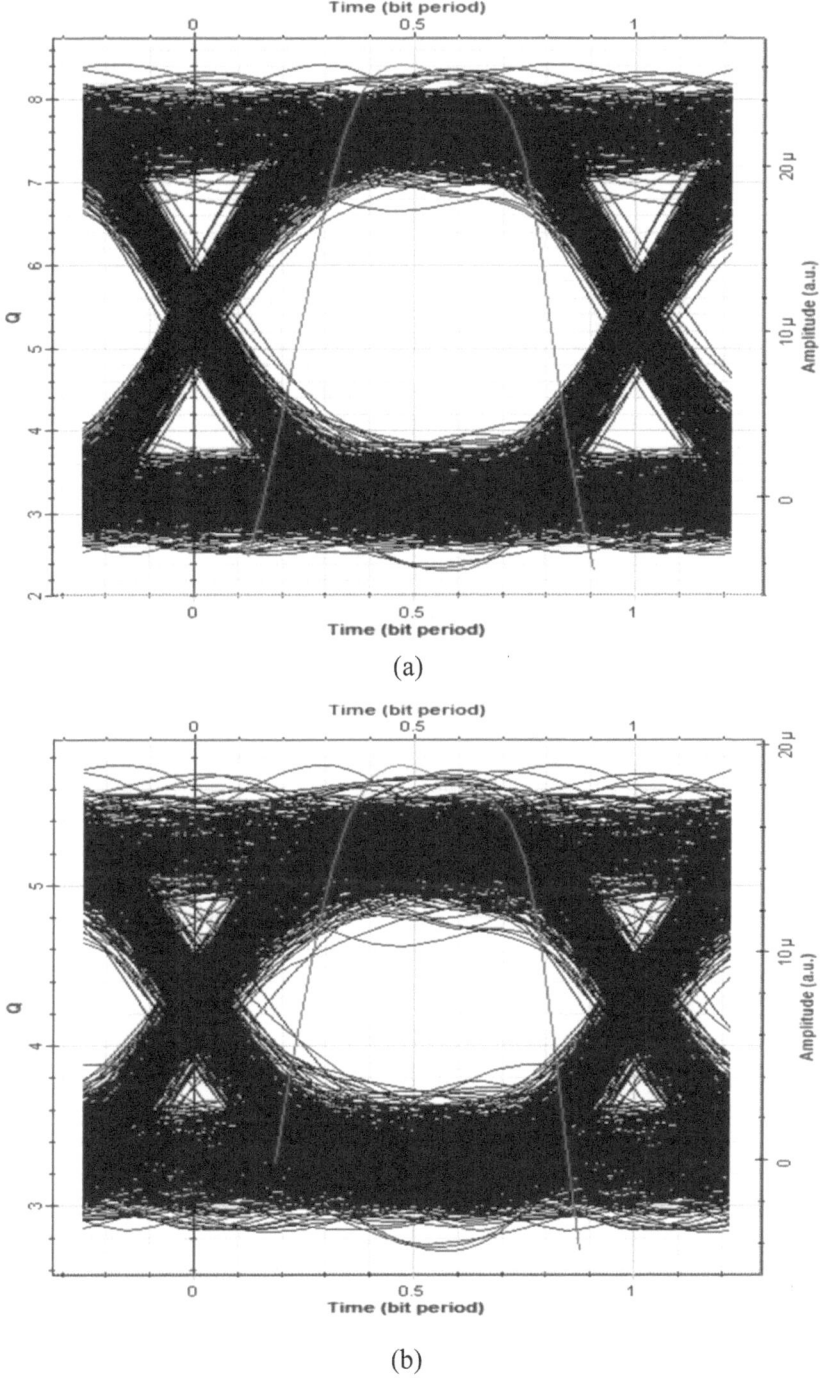

FIGURE 8.4 Eye diagrams of the received signal at 800 m under wet snow conditions for (a) LG00 mode and (b) LG01 mode.

mode channels, the Q factor of the received signal is determined as 14, 11 and 8 for LG00 mode and 10, 7 and 5 for LG01 mode over the FSO range of 700, 750 and 800 m, respectively. The results illustrate that with the rise in the transmission range, the performance of the system degrades.

Again, Figure 8.3 shows that over 700, 750 and 800 m of the FSO range, total received power is obtained as −86, −88 and −91 dBm for LG00 mode and −89, −90 and −94 dBm for LG01 mode because with increase in the transmission range, the power received decreases. Besides, signal to noise ratio (SNR) values are decreased with increase in the FSO range and 13, 12 and 8 dB for LG00 and 10, 7 and 5 dB for LG00 over the FSO range of 700, 750 and 800 m, respectively. Furthermore, the computed results in Figure 8.3 report that the performance of LG00 mode is considerably better than LG01 and thus LG00 performs the best by providing 40 Gbps transmission along the 800 m Ro-FSO range, followed by LG01 up to (740m) range. Figure 8.4 presents the eye diagrams of the recovered data signals at the user end at 800 m under wet snow conditions for LG00 and LG01 mode, respectively.

From the obtained results in Figure 8.5, it can be noted that the log of BER of the received signal for LG00 mode channel is computed as −52, −29 and −16 over the Ro-FSO range of 200, 210 and 220 m, respectively, at an input power of 10 dBm. Also, for LG01 mode channel, the log of BER of the received signal is determined as

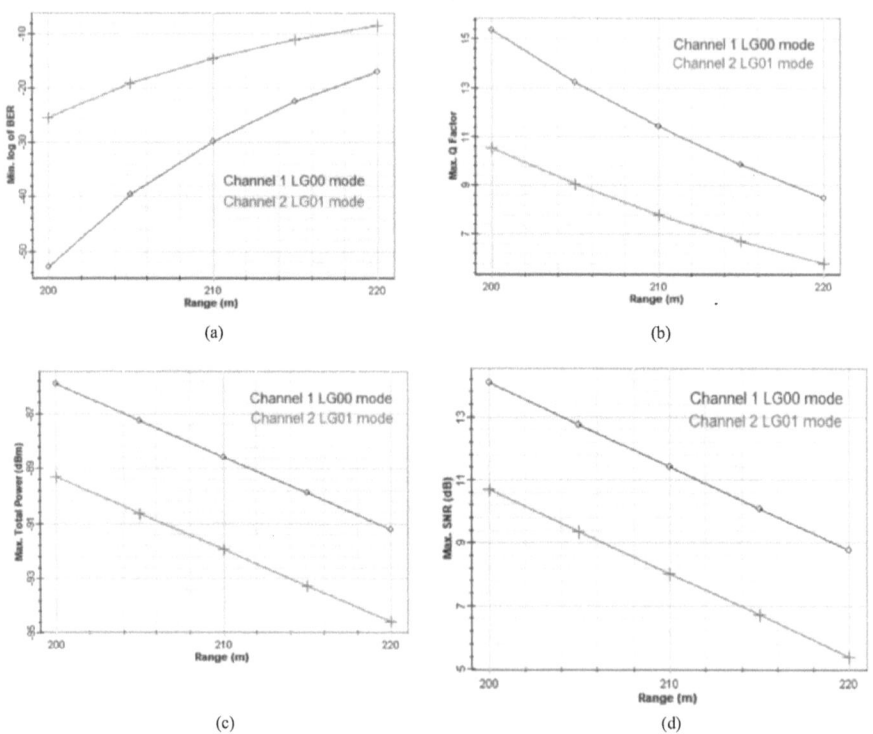

FIGURE 8.5   (a) Log of BER, (b) Q factor, (c) power and (d) SNR versus range under dry snow.

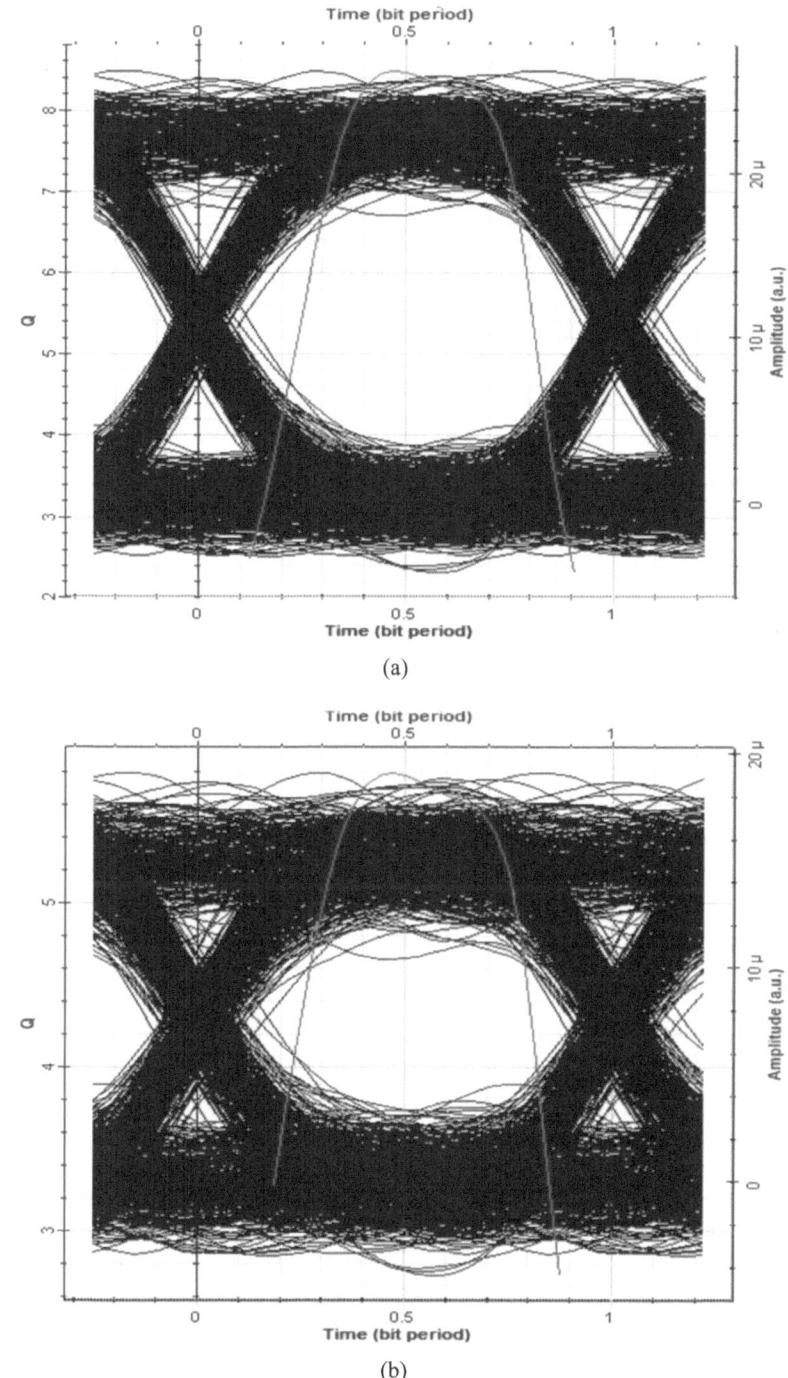

(a)

(b)

FIGURE 8.6 Eye diagrams of the received signal at 220 m under dry snow conditions for (a) LG00 mode and (b) LG01 mode.

−25, −14 and −8 over the FSO range of 200, 210 and 220 m, respectively. Similarly, as shown in Figure 8.5, at the same Ro-FSO range, for LG00 and LG01 mode channels the Q factor of the received signal is determined as 15, 11 and 8 for LG00 mode and 10, 7 and 5 for LG01 mode over the FSO range of 200, 210 and 220 m, respectively.

The results illustrate that with the rise in the transmission range, the performance of the system degrades.

Again, Figure 8.5 shows that over 200, 210 and 220 m of the FSO range, total received power is obtained as −85, −88 and −91 dBm for LG00 mode and −89, −91 and −94 dBm for LG01 mode because with increase in the transmission range, the power received decreases. Besides, SNR values are decreased with increases in the FSO range and 14, 11 and 8 dB for LG00 and 10, 8 and 5 dB for LG00 over the FSO range of 200, 210 and 220 m, respectively. Furthermore, the computed results in Figure 8.5 report that the performance of LG00 mode is considerably better than LG01 and thus LG00 performs the best by providing 40 Gbps transmission along the 220 m Ro-FSO range, followed by LG01 up to under dry snow. Figure 8.6 presents the eye diagrams of the recovered data signals at the user end at 220 m under wet snow conditions for LG00 and LG01 mode, respectively.

8.4 CONCLUSIONS

In this work, the performance of a designed 2 × 10 Gbps–10 GHz SDM-based Ro-FSO link over distinct Laguerre Gaussian modes (LG00 and LG01) under wet snow and dry snow in terms of BER, Q factor, received power and SNR is reported. The results present the successful 2 × 10 Gbps–10 GHz information transmission along 800 m link range under wet snow and 220 m link range under dry snow with acceptable Q Factor, BER, SNR, received power and eye diagrams. Also LG00 mode performs better than LG01 mode. In future, the proposed system can be used for high-speed backhaul transmission required for 5G networks and IoT applications.

REFERENCES

1. W.A. Imtiaz, H.Y. Ahmed, M. Zeghid, and Y. Sharief, "Two dimensional optimized enhanced multi diagonal code for OCDMA passive optical networks", *Opt. Quantum Electron.*, vol. 52, no. 1, pp. 1–17, 2020.
2. W. Ding et al., "A hybrid power line and visible light communication system for indoor hospital applications", *Comput. Ind.*, vol. 68, no. April 2018, pp. 170–178, 2015.
3. S. Chaudhary and A. Amphawan, "Selective excitation of LG 00, LG 01, and LG 02 modes by a solid core PCF based mode selector in MDM-Ro-FSO transmission systems", *Laser Phys.*, vol. 28, no. 7, p. aabd15, 2018.
4. P. Vijayakumari and M. Sumathi, "Physical implementation of underwater optical wireless system using spatial mode laser sources with optimization of spatial matching components", *Results Phys.*, vol. 14, p. 102503, 2019.
5. H. Maraha, K.A. Ameen, O.A. Mahmood, and A. Al-Dawoodi, "DWDM over FSO under the effect of different atmospheric attenuations", *Indones. J. Electr. Eng. Comput. Sci.*, vol. 18, no. 2, pp. 1089–1095, 2020.
6. A. Grover, A. Sheetal, and V. Dhasarathan, "20 Gbit/s-40 GHz OFDM based LEOGEO radio over inter-satellite optical wireless communication (Ro-IsOWC) system using 4-QAM modulation", *Optik (Stuttg).*, vol. 206, p. 164295, 2020.

7. J. Han, J. Zhang, Y. Zhao, and W. Gu, "Channel capacity and space-time block coding for coherent optical MIMO multi-mode fiber links", *Optik (Stuttg).*, vol. 124, no. 10, pp. 922–927, 2013.
8. A. Grover and A. Sheetal, "A cost-effective high-capacity OFDM based RoFSO transmission link incorporating hybrid SS-WDM-MDM of Hermite Gaussian modes", *Optoelectron. Adv. Mater. - RAPID Commun.*, vol. 14, no. 3–4, pp. 136–145, 2020.
9. C.Y. Lin et al., "A 400 Gbps/100 m free-space optical link", *Laser Phys. Lett.*, vol. 14, no. 2, p. aa548f, 2017.
10. C. Chen, W.-D. Zhong, and D. Wu, "Integration of variable-rate OWC with OFDMPON for hybrid optical access based on adaptive envelope modulation", *Opt. Commun.*, vol. 381, pp. 10–17, 2016.
11. A. Amphawan, Y. Fazea, and H. Ibrahim, "Mode division multiplexing of spiralphased donut modes in multimode fiber", *Int. Conf. Opt. Photonic Eng. (icOPEN 2015)*, vol. 9524, no. icOPEN, p. 95240S, 2015.
12. M. Singh and J. Malhotra, "Long-reach high-capacity hybrid MDM-OFDM-FSO transmission link under the effect of atmospheric turbulence", *Wirel. Pers. Commun.*, vol. 107, no. 4, pp. 1549–1571, 2019.
13. S. Chaudhary et al., "40 Gbps–80 GHz PSK-MDM based Ro-FSO transmission system", *Opt. Quantum Electron.*, vol. 50, no. 8, pp. 1–9, 2018.
14. A. Grover, A. Sheetal, and V. Dhasarathan, "Enhanced performance analysis of a hybrid spectrum sliced-wavelength division multiplexing-mode division multiplexing orthogonal frequency division multiplexing based radio-over-free space optics", *Trans. Emerg. Telecommun. Technol.*, vol. 31, no. 6, pp. 1–11, 2020.
15. M.N. Ismael, A. Al-Dawoodi, S. Alshwani, A. Ghazi, and A.M. Fakhrudeen, "SDM over hybrid FSO link under different weather conditions and FTTH based on electrical equalization", *Int. J. Civ. Eng. Technol.*, vol. 10, no. 1, pp. 1396–1406, 2019.
16. S. Alshwani et al., "Hermite-Gaussian mode in spatial division multiplexing over FSO system under different weather condition based on linear Gaussian filter", *Int. J. Mech. Eng. Technol.*, vol. 10, no. 1, pp. 1095–1105, 2019.
17. A. Al-Dawoodi et al., "Investigation of 8×5 GB/S mode division multiplexing-FSO system under different weather condition", *J. Eng. Sci. Technol.*, vol. 14, no. 2, pp. 674–681, 2019.
18. M. Singh, J. Malhotra, M.S. Mani Rajan, D. Vigneswaran, and M.H. Aly, "A longhaul 100 Gbps hybrid PDM/CO-OFDM FSO transmission system: Impact of climate conditions and atmospheric turbulence", *Alexandria Eng. J.*, vol. 60, no. 1, pp. 785–794, 2020.
19. V. Dhasarathan, M. Singh, and J. Malhotra, "Development of high-speed FSO transmission link for the implementation of 5G and Internet of Things", *Wirel. Networks*, vol. 26, no. 4, pp. 2403–2412, 2020.
20. M.A. Esmail, A. Ragheb, H. Fathallah, and M.S. Alouini, "Investigation and demonstration of high speed full-optical hybrid FSO/fiber communication system under light sand storm condition", *IEEE Photonics J.*, vol. 9, no. 1, 2017, pp. 1–12.
21. M.N. Ismae, A. Al-Dawoodi, and A.A. Mohammed, "Hybrid coarse and dense WDM over FSO link under the effect of moderate and heavy rain weather attenuations", *J. Eng. Sci. Technol.*, vol. 15, no. 5, pp. 3494–3501, 2020.
22. A. Amphawan, S. Chaudhary, R. Din, and M.N. Omar, "5 Gbps HG01 and HG03 optical mode division multiplexing for RoFSO", *Proceedings on 2015 IEEE 11th Int. Colloq. Signal Process. Its Appl. CSPA* 2015, pp. 145–149, 2015.
23. K. Pang et al., "400-Gbit/s QPSK free-space optical communication link based on fourfold multiplexing of Hermite–Gaussian or Laguerre–Gaussian modes by varying both modal indices", *Opt. Lett.*, vol. 43, no. 16, p. 3889, 2018.
24. M. Balasaraswathi, M. Singh, J. Malhotra, and V. Dhasarathan, "A high-speed radioover-free-space optics link using wavelength division multiplexing-mode division multiplexing-multibeam technique", *Comput. Electr. Eng.*, vol. 87, p. 106779, 2020.

25. Z. Zhou, E. Li, and H. Zhang, "Performance analysis of duobinary and AMI techniques using LG modes in hybrid MDM-WDM-FSO transmission system", *J. Opt. Commun.*, pp. 1–7, 2019.

26. S. Chaudhary, A. Amphawan, N. Ali, and M.S. Sajat, "4X2.5 Gbps-10 GHz WDMMDM Ro-FSO with Spiral-Phased LG-HG Modes", *4th International Conference on Internet Applications, Protocols and Services (NETAPPS2015)*, pp. 241–244, 2015.

27. V. Sharma and A. Kaur, "Optimization of Inter-satellite Link (ISL) in Hybrid OFDM-IsOWC Transmission System", pp. 3–6, 2013.

28. P. Sharma, "Analysis of pointing error at 100 Gbps by using different encoding technique in spatial diversity based inter-satellite optical wireless communication (IS-OWC) system", *Int. J. Eng. Res. Technol.*, vol. 7, no. 09, pp. 261–264, 2018.

29. M. Singh, J. Malhotra, and V. Dhasarathan, "A high-speed single-channel intersatellite optical wireless communication link incorporating spectrum-efficient orthogonal modulation scheme", *Microw. Opt. Technol. Lett.*, vol. 62, no. 12, pp. 4007–4014, 2020.

30. S. Chaudhary, X. Tang, A. Sharma, B. Lin, X. Wei, and A. Parmar, "A cost-effective 100 Gbps SAC-OCDMA–PDM based inter-satellite communication link", *Opt. Quantum Electron.*, vol. 51, no. 5, pp. 1–10, 2019.

9 Optimization of Energy Consumption in the Wireless Sensor Network (WSN) Using IoT-Based Cross-Layer Model

Shobhit Tyagi, Abhishek Kumar, and Vishwas Mishra
Swami Vivekananda Subharti University

Amita Soni
Christ University

Ramesh Chandra Rath
GGSESTC

Shyam Akashe
ITM University

CONTENTS

9.1 INTRODUCTION

These days, a WSN could be a gathering of autonomous hubs or terminals that speak with one another by shaping a multi-bounce specially appointed system. Such WSNs could change their topology progressively because of hubs portability. WSNs have a few confinements, for example, constrained vitality supply, restricted figuring force, and restricted transmission capacity. Vitality effectiveness has been viewed as a standout among the most significant structure difficulties in WSN. In the layered methodology, the exchange overhead is more which devour more vitality. Because of these confinements, cross-layer configuration is developed for WSNs. Cross-layer conventions don't pursue the layered structure as a customary OSI model pursues. In the cross-layered methodology, the convention stack is treated as a framework and not an autonomous individual layer. The cross-layer approach for WSNs is more powerful than the customary methodology. Cross-layer approach expresses that data of at least two layers are utilized to accomplish an ideal goal. This chapter introduces a cross-layer operation model that can improve the energy consumption and system throughput of IEEE 802.15.4 MWSNs. The proposed model integrates four layers in the network operation: (1) application (node location); (2) network (routing); (3) medium access control (MAC); and (4) physical layers. The location of the mobile nodes is embedded in the routing operation after the route discovery process. The location information is then utilized by the MAC layer transmission power control to adjust the transmission range of the node. This is used to minimize the power utilized by the network interface to reduce the energy consumption of the node(s). The model employs a mechanism to minimize the neighbor discovery broadcasts to the active routes only. Reducing control packet broadcasts between the nodes reduces the network's consumed energy. It also decreases the occupation period of the wireless channel. The model operation leads the network to consume less energy while maintaining the network packet delivery ratio, energy efficiency, throughput improvement, and low delay [1,2].

The study of cross-layer optimization is a relatively new research area. Cross-layer design across the various layers of the protocol stack poses many difficult and challenging problems. An important question in the area of cross-layer design is what parameters need to be shared among different layers of the protocol stack and how can each layer be made robust to the changing network conditions.

The benefits and advantages from relaxing the rigid layered structure needs to be quantified, and the associated complexity and stability issues with implementing such cross-layer design need to be studied more thoroughly. Hence, it will be essential to fully understand the impact of the outputs from one layer on the functions of other layers. The parameters in a particular layer that shaped the functions in other layers need to be identified, and a design methodology for such a cross-layer approach that exploits the inter-layer relationships also needs to be developed [3].

Cross-layer design in wireless networks is strongly recommended where the characteristics of the wireless channel permeate the functions of all protocol layers in the traditional protocol stack. Other characteristics of the wireless environment, such as mobility, fading, and interferences, also introduce a strong inter-layer relationship and their effects need to be fully understood and considered in the network design [4].

There are numerous advantages for implementing a cross-layer design and optimization technique. To name one, the technique can help improve efficiency at both the node and system level by improving the efficiency of protocol functions that utilize critical information from other protocol layers and thus creating a more optimized performance at each protocol layer. For example, this technique can help increase a node or network capacity and make it more flexible and robust to bandwidth variations and more tolerant to data losses since data transmission is dependent on the state of channel conditions and data losses may occur frequently in wireless networks [5]. It can also help lengthen the lifetime of wireless devices by improving the power efficiency in portable wireless devices, such as handheld terminals in cellular networks or sensor nodes in WSNs. The cross-layer design and optimization technique can also enhance the capability of the network to support a wider range of applications; for example, in the context of WSNs, a sensor node may support multiple functions for both sensing and ad-hoc operations.

In this chapter, a cross-layer optimization framework is proposed. It allows a systematic and holistic approach to study the interactions between various different protocol layers in any wireless network. The structure of the optimization agent in the proposed framework provides a modular and scalable framework when changes or modifications need to be made to accommodate different requirements or applications for various different types of networks and the benefits of such study will help to provide valuable insights into the design of next-generation algorithms and protocols for wireless systems, especially for WSNs.

The energy model of the remote sensor system can be characterized as the complete vitality utilization of the system, including every one of its units, be it sensor gadget parts, vitality devoured in directing or course support, topology upkeep or whosoever it might be. Creating a vitality model is a significant piece of any convention advancement and its presentation assessment. Here, we considered a system with n portable sensor hubs and one sink hub which is static. Along these lines, vitality model can be portrayed as beneath: Energy devoured by sensor gadget: the sensor gadget contains preparing units, detecting unit, memory unit, and handset unit. Thus, vitality utilization of every unit should be considered [6–8].

$$E_{\text{Node}} = E_{\text{controller}} + E_{\text{sensor}} + E_{\text{memory}} + E_{\text{Transreceiver}} \tag{9.1}$$

where E_{Node} is the total energy consumed by a sensor device, $E_{\text{controller}}$ is the energy consumed by the controlling units, E_{sensor} is the energy consumed by the sensing unit, E_{memory} is the energy consumed by the memory unit, and $E_{\text{Transreceiver}}$ is the energy consumed by the transceiver unit.

9.2 CROSS-LAYER MODELING OF WSN

The layered approach has many disadvantages when compared to the cross-layer approach. The cross-layered approach improves network (Figure 9.1).

Software architecture of the proposed model is shown in Figure 9.2. Application layer, network layer, data link layer, and Internet of Things (IoT) physical layer are involved in the operational model. The MAC layer is included in data link layer [9]. The practical deployment of cross-layer scheme in the proposed transmission model of sensor networks is an important task. We should not only design MAC layer protocol with network layer location information but also take PHY layer conditions into consideration, for example, the diversity reception of path loss transmission and the anti-jamming for transmission security and full duplex technique in PHY layer. The appropriate PHY IoT layer propagation models for different transmission

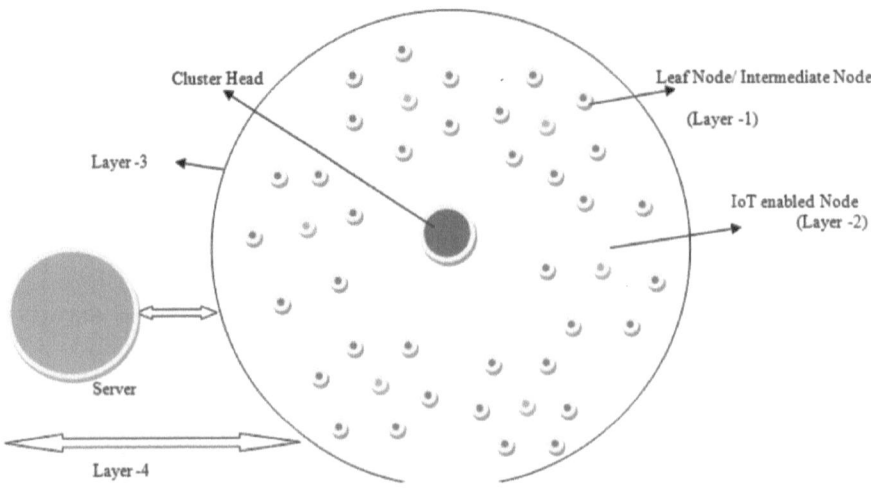

FIGURE 9.1   Architecture of wireless sensor network.

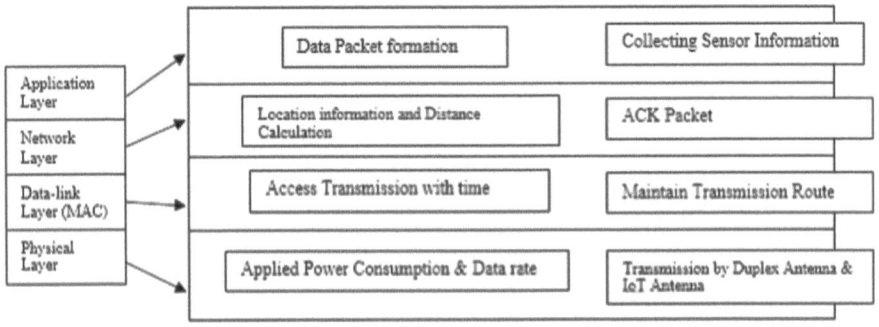

FIGURE 9.2   Software architecture of proposed work.

environments are necessary to the practical deployment of sensor networks. The PHY layer propagation models are also needed to support upper layers to complete the whole transmission. Application layer collects information from environment. In this research work an IoT-based four layered cross-layer model for mobile wireless sensor network (WSN) proposed to design and analyze.

In our transmission model for versatile remote sensor organize, we expect that all center points in the structure are in Poisson dissemination. There are two sorts of full-duplex transmission course in the proposed model. One is the bidirectional course, which suggests that center point A transmits information to center B and center point B transmits information to center point meanwhile. The other is the unidirectional course, where center point A transmits information to center B and center B transmits information to center C. Each center point in the model has two sorts of transmission go, as shown in Figure 9.3. One is Data Transmission Range (DTR). The other is Sensing Transmission Range (STR). In a center point's DTR,

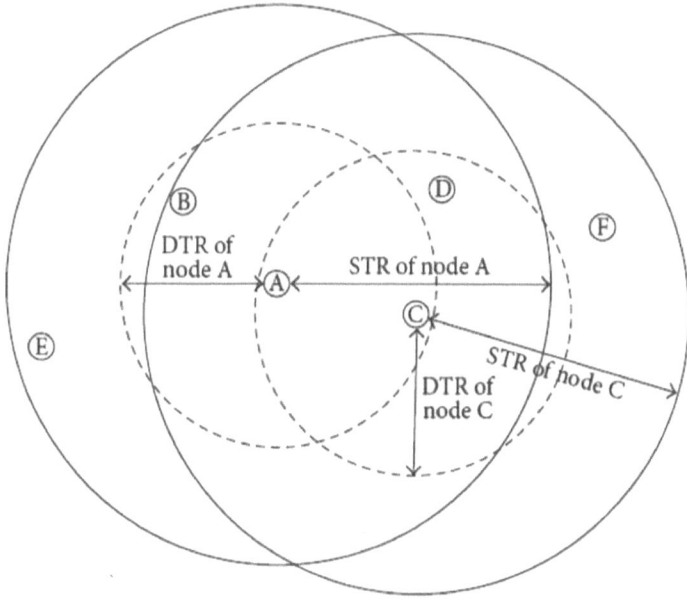

FIGURE 9.3   Definition of DTR and STR.

it can successfully send data to objective center. In a center's STR, it can recognize the transmission of all sending center points in the locale. In this chapter, we expect that both the STRs and the DTRs of the center points are equal. The importance of different ranges can deal with covered center point issues. As shown in Figure 9.3, center F is in the STR of center C yet not of center A, which may go about as a covered center in the transmission between centers An and C. Along these lines, in bidirectional course transmission, covered center points issue can be settled by full-duplex interface in light of simultaneous transmission. In unidirectional course transmission, center point F can distinguish the advancing transmission from center point C by RTS/CTS instrument [10]. From this time forward, the issue is disentangled.

9.3   PROBLEM FORMULATION

Energy consumption in system instatement [2]: At the season of system introduction, hubs need to discover their neighbors. Assurance of the absolute number of neighbors additionally plays an incredible effect on the general vitality utilization. In this way, for discovering neighbors, hubs trade neighbor revelation (ND) parcels. Along these lines, vitality devoured in system introduction is the aggregate of the vitality expended in sending and getting ND packets.

$$E_{\text{init}} = \sum_{i=1}^{n} E_{\text{ndp}} \tag{9.2}$$

where n is the number of nodes, E_{init} is the total energy consumed in initialization phase and E_{ndp} is the energy consumed in exchanging neighbor discovery packets.

Energy consumption in course disclosure: For course revelation, source hubs communicate the RQ parcels and in the wake of accepting the RQ bundle, the sink hub unicasts RP bundle. Here RQ bundles were permitted to be communicated by $x/3$ hubs just, i.e., the main $x/3$ hubs whose weight is most astounding among every one of the hubs present in the neighbor table. Rather than broadcasting RQ bundles by every one of the hubs, just $x/3$ hubs communicate the parcel. Along these lines, vitality utilization is limited. Alongside RQ parcels, hubs intermittently trade ACK bundles for neighbors' support. Along these lines, Energy expended in this stage is the absolute vitality utilization in communicating RQ parcels in addition to trading ACK bundles.

$$E_{P1} = \sum_{i=1}^{x/3} E_{RQ} + \sum_{i=1}^{n} E_{ACK} \tag{9.3}$$

where E_{P1} is the energy consumed in phase 1, E_{RQ} is the energy consumed in transmitting RQ packets and E_{ACK} is the energy consumed in exchanging ACK packets. In the wake of accepting the RQ bundle from the goal hub, the RP parcel is casted. As RP bundle is casted, it makes another arrangement of hubs for example 'm'. We likewise limit the intermittent trade of hi messages with the hubs present in the dynamic course. Hence, vitality utilization at this stage is the absolute vitality devoured in casting RP bundle in addition to trading an ACK parcel.

$$E_{P2} = \sum_{i=1}^{m} E_{RP} + \sum_{i=1}^{m} E_{ACK} \tag{9.4}$$

where E_{P2} the energy is consumed in phase 2, E_{RP} is the energy consumed in transmitting RP packets and E_{ACK} is the energy consumed in exchanging ACK packets.

Energy consumption in information transmission: After the foundation of the course, the hub transmits information to the goal or sink hub. In this way, the vitality utilization here is the aggregate of the vitality expended in transmission of the information bundles and the ack parcels.

$$E_{dat} = \sum_{i=1}^{m} E_d + \sum_{i=1}^{m} E_{ACK} \tag{9.5}$$

where m is the number of nodes participating in route formation, E_{dat} is the total energy consumed in transmitting data, E_d is the energy consumed in transmitting data and E_{ACK} is the energy consumed in exchanging ACK packets.

Energy consumption by sink: Energy utilization at the sink is the absolute vitality devoured by the source hubs which speaks with the sink and plays out the sink directions.

$$E_{sink} = \sum_{i=1}^{s} E_{source} \tag{9.6}$$

where s is the number of source node, E_{sink} is the total energy consumed by sink, E_{source} is the energy consumed by the source nodes. A. Energy consumption by other factors: A node can be in a sleep mode at some point of time, so the state transition from run to sleep, sleep to idle and idle to run also consume energy. Idle listening and overhearing also results in the consumption of energy. So, this factor also needs to be considered in energy model.

$$E_{loss} = E_{Tidle} + E_{overhear} + E_{tr} + E_{sleep} \qquad (9.7)$$

where E_{loss} is the total energy consumed by loss factors, E_{Tidle} is the energy consumed when sensor is idle, $E_{overhear}$ is the energy consumed in overhearing, E_{tr} is the energy consumed in state transition and E_{sleep} is the energy consumed when the sensor is in sleep mode. Hence, the total energy consumption of the network is given as

$$E_{total} = E_{init} + E_{sensor} + E_{p1} + E_{p2} + E_{dat} + E_{sink} + E_{loss} \qquad (9.8)$$

where E_{total} is the total energy consumed by considering all factors. The proposed model is evaluated on the following parameters:

1. **Total network energy consumption**: Total network energy consumption is represented by the sum of energy consumed by each node of the network in transmitting data, finding neighbors and routes etc.

$$E_{net} = \text{Total Energy Consumed due to all Motes} \qquad (9.9)$$

2. **Energy per packet**: The energy per packet metric is calculated as the ratio of energy consumed by the network to the total number of packets transmitted by the source.

$$E_{packet} = \frac{E_{net}}{\text{Packets Transmitted by Source}} \qquad (9.10)$$

3. **Throughput**: Throughput is the rate of the amount of data delivered from source to destination during a unit of time.

$$Th = \frac{\text{Number of Packets Received}}{\text{Network Operation time}} \qquad (9.11)$$

4. **Packet delivery Ratio (PDR)**: It is defined as the ratio of the number of data packets delivered to the destination.

$$PDR = \frac{\text{No. of Packets Received}}{\text{No. of Packets Transmitted}} \qquad (9.12)$$

5. **Control overhead**: It is the total number of control messages broadcasted in the entire networks.

6. **Normalized routing load (NRL)**: It is determined as the proportion of an all out number of control parcels sent to the absolute number of information bundles gotten by the goal.

$$\text{Routing Network Load} = \frac{\text{No. of Packets Send}}{\text{No. of packets Received}} \qquad (9.13)$$

7. **End-to-end delay**: It is calculated as the time taken for a packet to be sent across the network, i.e., from source to destination.

$$T_d = \frac{\text{Delay in every Successful packet Sent}}{\text{Total No. of Packets Generated}} \qquad (9.14)$$

8. **Fair Index**: to evaluate the fairness of nodes accessing the channel and it indicates the fairness of nodes accessing the networks. FI is presented as

$$FI = \frac{\left(\sum_i Th_i\right)^2}{n \sum_i Th_i^2} \qquad (9.15)$$

where n is the number of contending nodes and Th_i is the throughput of node i [11,12].

9.4 RESULTS

The simulation is performed in MATLAB software and simulation parameters of the IoT-enabled cross-layer WSN are shown in Table 9.1.

TABLE 9.1
Wireless Sensor Network Parameters

S. No.	Parameter	Value
1	Node battery energy E_b	1,000 kJ
2	E_{dm}	9.37 µJ
3	E_{rts}	1.45 µJ
4	E_{cts}	2.86 µJ
5	E_{rrp}	3.87 µJ
6	E_{ack}	0.472 µJ
7	$E_{datatransfer}$	4 mJ
8	E_{hello}	120 µj
9	Area A	$300 \times 300 \, m^2$
10	Number of nodes	200
11	Maximum queue length	150
12	Number of bits per packet	1,024 bis

Simulated results are shown in figure below. Figure 9.4 shows comparison result of energy consumption for normal cross-layer and the IoT-based cross-layer models. It can be observed that for the normal cross-layer model for 20 nodes energy consumption is 12 kJ while that for the proposed model is 11.6 kJ. And for 200 nodes normal cross-layer model, it consumes 34 kJ while proposed model consumption is 30 kJ.

Figure 9.5 shows comparison result of delay for normal cross-layer and IoT-based cross-layer model. It can be observed that for normal cross-layer model for 20 nodes of delay is 0.1 s while that for the proposed model is 0.095. And for 200 nodes normal cross-layer model takes 0.98 s for transmission while the proposed model takes 0.94 s.

Figure 9.6 shows comparison result of percentage of packet delivery ratio for normal cross-layer and IoT-based cross-layer model. It can be observed that for normal cross-layer model for 20 nodes of PDR is 60% while that for the proposed model is 70%. And for 200 nodes normal cross-layer model PDR is 92% for transmission while the proposed model PDR is 97%.

Figure 9.7 shows comparison result of Throughput for normal cross-layer and IoT-based cross-layer model. It can be observed that for normal cross-layer model for 20 nodes of Throughput is 0.1 Mbps while that for the proposed model is 0.11 Mbps. And for 200 nodes normal cross-layer model Throughput is 4.2 Mbps for transmission while the proposed model Throughput is 4.5 Mbps.

Figure 9.8 shows comparison result of Fair index for normal cross-layer and IoT-based cross-layer model. It can be observed that for normal cross-layer model for 20

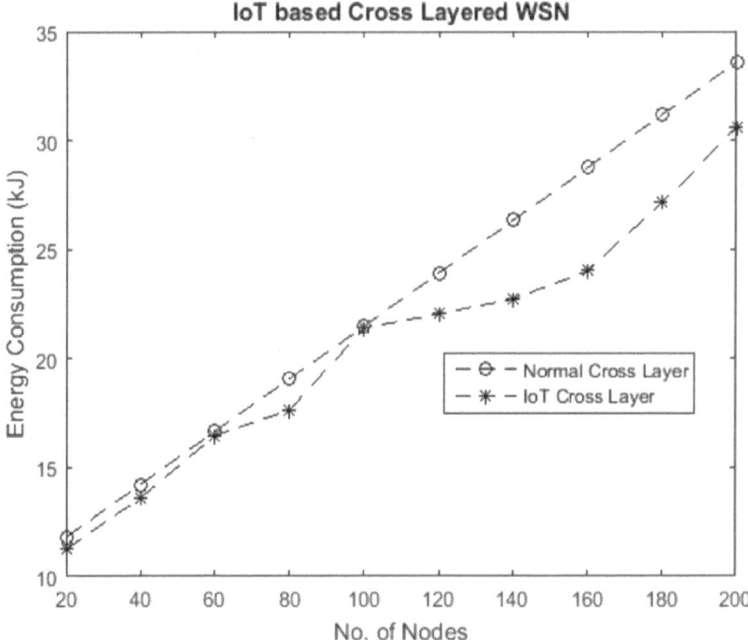

FIGURE 9.4 Comparison of energy consumption for cross layer and IoT-enabled four layered cross-layer model.

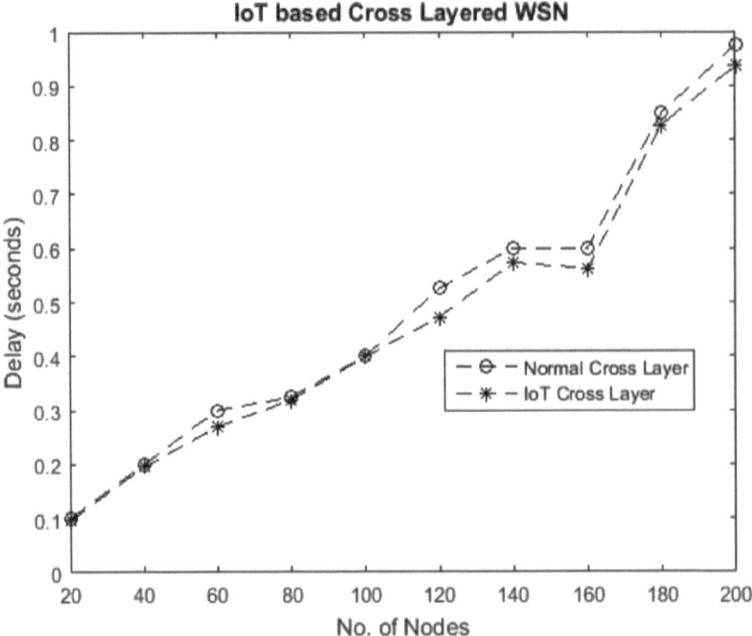

FIGURE 9.5   Comparison of delay for cross layer and IoT-enabled four layered cross-layer model.

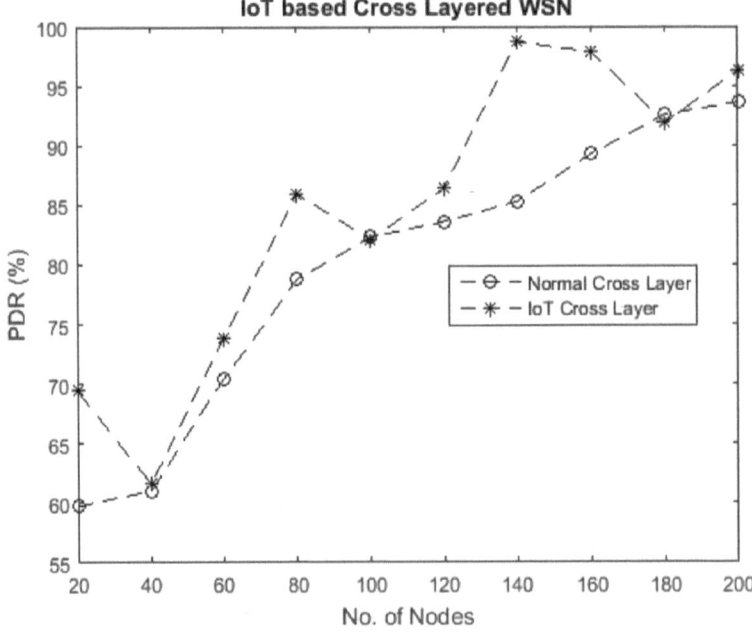

FIGURE 9.6   Comparison of PDR for cross layer and IoT-enabled four layered cross-layer model.

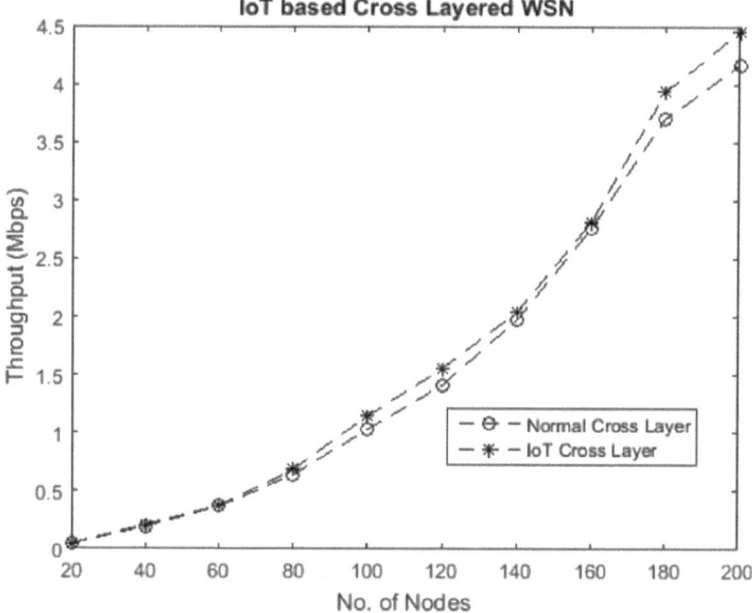

FIGURE 9.7 Comparison of throughput for cross layer and IoT-enabled four layered cross-layer model.

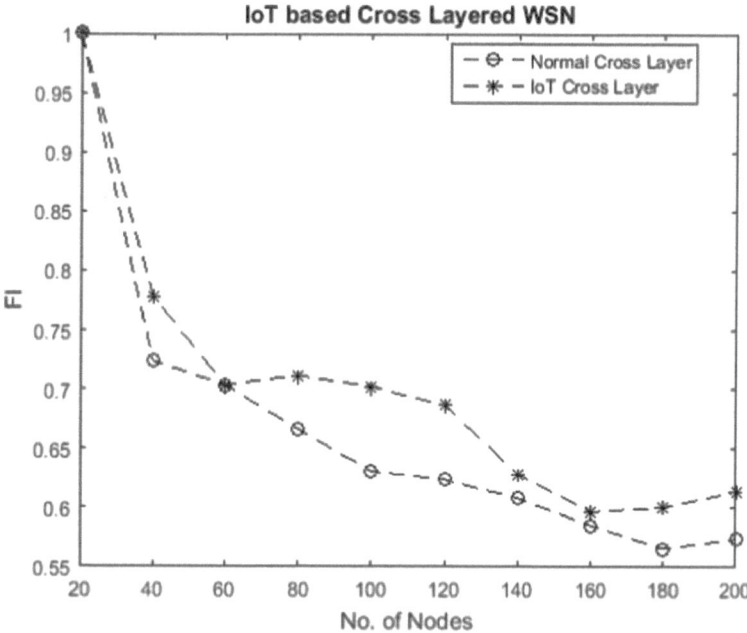

FIGURE 9.8 Comparison of fair index for cross layer and IoT-enabled four layered cross-layer model.

nodes of FI is 0.99 while that for the proposed model is 0.99. And for 200 nodes normal cross-layer model FI is 0.57 for transmission while the proposed model FI is 0.62.

9.5 CONCLUSIONS

A cross-layer model for the mobile WSN is designed and simulated in this research. The findings indicate that the proposed cross-layer method worked much better using a good mix of the network layer, data link layer, application layer, and IoT deployed cross layer. Its performance is carried out in the form of five energy consumption, transmission delay, packet distribution ratio (PDR), throughput and equal index (FI) parameters simulated on the MATLAB software. Here, it is concluded that the performance of the cross-layer model of WSN based on the IoT is 20% higher than the conventional cross-layer model. The simulation is performed and illustrated with MATLAB software and its plotting illustrates a comparison of conventional cross-layer design and IoT-based cross-layer design, and the results of the IoT-based cross-layer model are so good. Further enhancement in IoT-based cross-layer modeling can be implemented by using sleep-aware scheduling. Furthermore, the integration of the security algorithm when developing the WSN provides a wide variety of flawless data transfer across the network. Thus, the constant iteration of advanced algorithms contributes to the emergence of new approaches.

REFERENCES

1. Ball J.C., Ross A. (1991) *The Effectiveness of Methadone Maintenance Treatment: Patients, Programs, Services and Outcomes.* Springer, Berlin Heidelberg, New York.
2. Jellinek E.M. (1960) *The Disease Concept of Alcoholism.* Hillhouse, New Haven.
3. Snider T., Grand L. (1982) *Air Pollution by Nitrogen Oxides.* Elsevier, Amsterdam.
4. Noller C., Smith V.R. (1987) Ultraviolet selection pressure on earliest organisms. In: Kingston H., Fulling C.P. (eds.) *Natural Environment Background Analysis.* Oxford University Press, Oxford, pp. 211–219.
5. Treasure J., Holland A.J. (1989) Genetic vulnerability to eating disorders: Evidence from twin and family studies. In: Remschmidt H., Schmidt M.H. (eds.) *Child and Youth Psychiatry: European Perspectives.* Hogrefe and Hubert, New York, pp. 59–68.
6. Zippel J., Harding F.W., Lagrange M. (1992) The stress of playing God. In: Mildor E. (ed.) *Explorations in Geopolitics*, 4th edn. Wiley, New York, pp. 103–124.
7. Arkhipenko D.K., Bokiy G.B. (1986) Factor-groups analysis and X-ray study of vermiculite and talc cristals (in Russian). *J Struct Chem* 16: 450–457.
8. Peters S., Jaffe H.G. (1991) Lactose synthesis and the pentose cycle. *J Biol Chem* 98: 15–33.
9. Uno H., Tarara R., Else J.G., Suleman M.A., Sapolsky R.M. (1989) Hippocampal damage associated with prolonged and fatal stress in primates. *J Neurosci* 9: 1705–1711.
10. Boshoff G.A. (1999) Development of integrated biological processing for the biodesalination of sulphate- and metal-rich wastewaters. Ph.D. thesis, Rhodes University.
11. Dyrynda P.E.J. (1996) An appraisal of the early impacts of the 'Sea Empress' oil spill on shore ecology within South-West Wales. (Final report by the University of Wales, Swansea for the Wildlife Trusts and WWF-UK).
12. Saavedra C. (1989) Manu – two decades later. WWF report June/July. WWF, Gland, pp. 6–9.

10 Study and Analysis of Average Throughput Based on the Performance of Broadcast Scheme in Wireless Ad-hoc Networks

Sanjeev Kumar Srivastava
Pillai College of Engineering

CONTENTS

10.1 INTRODUCTION

An ad-hoc wireless network is a network that is composed of individual devices communicating with each other directly. Such devices can communicate with another node that is immediately within their radio range or one that is outside their radio range. An ad-hoc wireless network is self-organizing and adaptive. Ad-hoc nodes or devices should be able to detect the presence of other such devices and to perform the necessary handshaking process to allow communications and the sharing of information and services.

In the scenario of broadcast scheme in ad-hoc wireless networks, it has been discussed briefly about hidden node problem in unicast and broadcast communication

DOI: 10.1201/9781003143802-10

in order to compare slotted ALOHA and threshold conditions based on the MAC layer. In the single-hop scheme, a mathematical model has been developed and analyzed to achieve throughput capacity in ad-hoc wireless networks. In the multi-hop scheme, an analytical model is presented for the computation of the throughput of a node network based on the IEEE 802.11 DCF mode under ideal channel conditions. In the simulation model, it has been assumed that the practical scenario is possible to check the validity of comparison. Several physical layer parameters as per the IEEE standard are also assumed.

10.2   NETWORK MODEL

10.2.1   SINGLE-HOP SCHEME

In this scheme, **node A**, which is in receiving range of **node B** but not in the receiving region of **node C**, may cause hidden terminal problem, as shown in Figure 10.1. This area is defined as potential hidden node area. For unicast communications, the size of the potential hidden node area is calculated using the distance between the sender and receiver. In broadcast communication, the potential hidden node area depends on the receiving range of all the neighboring nodes. Therefore, it is difficult to exactly compute the size of this area. By varying the carrier sensing area, the form of this area is also changed. If R denotes the transmission range of a node, the maximum size of potential hidden node area can be $(2R)^2 - \pi R^2 = 3\pi R^2$.

Thus, in the case of broadcast, the potential hidden terminal area is larger than that of unicast.

To do so, there are some assumptions which are as follows:

1. All nodes in the networks are two-dimensionally Poisson distributed with density λ, i.e., the probability $p(i, A)$ of finding i nodes in an area of size A is given by,

$$p(i, A) = \frac{(\lambda A)^i e^{-\lambda A}}{i!} \tag{10.1}$$

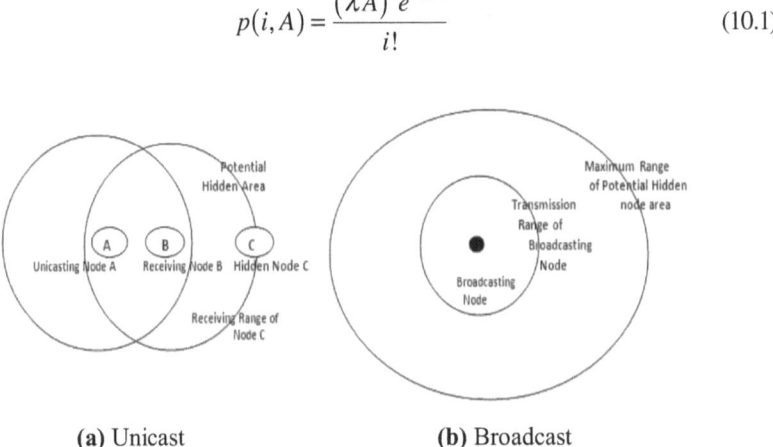

(a) Unicast   (b) Broadcast

FIGURE 10.1   Potential hidden nodes area.

2. All nodes have the same uniform transmission and receiving range of radius R. N is the average number of neighbor nodes within a circular region of πR^2. Therefore,

$$N = \lambda \pi R^2 \qquad (10.2)$$

3. A node transmits a frame only at the beginning of each slot time; however, the threshold condition-based MAC protocol node transmits only if the minimum threshold condition gets satisfied. The size of a slot time, τ, is the duration including transmit-to-receive turn-around time, carrier sensing delay and processing time.
4. The transmission time or the frame length is the same for all nodes, i.e., the same data packet length.
5. When a node is transmitting, it cannot receive at the same time.
6. A node is ready to transmit with probability p; however, for the MAC protocol, transmitting probability p also depends on the node's buffered data size. Let p' denote probability that a node transmits in a time slot. If p' is independent at any time slot, it can be obtained by,
7. $p' = p.\text{Prob}\{\text{channel is sensed idle in } a \text{ slot}\} \approx p.P_I.$
   Where P_I is the limiting probability that the channel is sensed to be idle.
8. The carrier sensing range is assumed to vary between the range $R \sim 2R$.

From the above-mentioned assumptions, the channel process modeled can be represented as a two-state Markov chain, as shown in Figure 10.2.

This model has two states and is described as follows:

Idle: This is the state when the channel around node 'x' is sensed idle, and its duration T_{idle} is τ

Busy: This is the state when a successful DATA transfer is done. The channel is in busy state for the duration of DATA transfer, thus the busy time, T_{busy}, is equal to the data transmission time

$$\delta_{\text{data}} \cdot \left(T_{\text{busy}} = \delta_{\text{data}} \right) \qquad (10.3)$$

In the MAC scheme, all nodes should stay at least ideal for one slot time, after the channel becomes idle. Thus, the transition probability P_{bi} is 1.

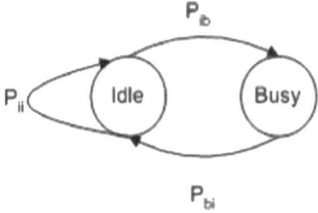

FIGURE 10.2   Markov chain model for the channel.

P_{ii} is the transition probability of the neighbor nodes transmission and is given by

$$P_{ii} = \sum_{i=0}^{\infty} (1-p')^i \frac{(\lambda \pi R^2)^i}{i!} e^{-\lambda \pi R^2} = e^{-p'N} \tag{10.4}$$

Let, Φ_i and Φ_b denote the steady-state probabilities of idle and busy states, respectively.

$$\Phi_i = \Phi_i P_{ii} + \Phi_b P_{bi} = \Phi_i P_{ii} + \Phi_b \tag{10.5}$$

$$\text{Since } \Phi_b = 1 - \Phi_i, \Phi_i = \frac{1}{2 - P_{ii}} = \frac{1}{2 - e^{-p'N}} \tag{10.6}$$

Now the limiting probability P_l can be obtained by

$$P_l = \frac{\tau}{(\delta_{data})(1 - e^{-p'N}) + \tau} \tag{10.7}$$

According to the relationship between P' and P, P' is given by

$$P' = \frac{\tau p}{(\delta_{data})(1 - e^{-p'N}) + \tau} \tag{10.8}$$

For the throughput calculation, it needs to calculate the probability of a successful transmission.

As shown in Figure 10.3, the transmission state of node 'x' can be modeled by the three states Markov chain model. Wait, succeed and collision states represent the node's transmission difference, successful DATA transmission, and collision state conditions, respectively. At the beginning of each time slot, node 'x' leaves the wait sate with probability p'.

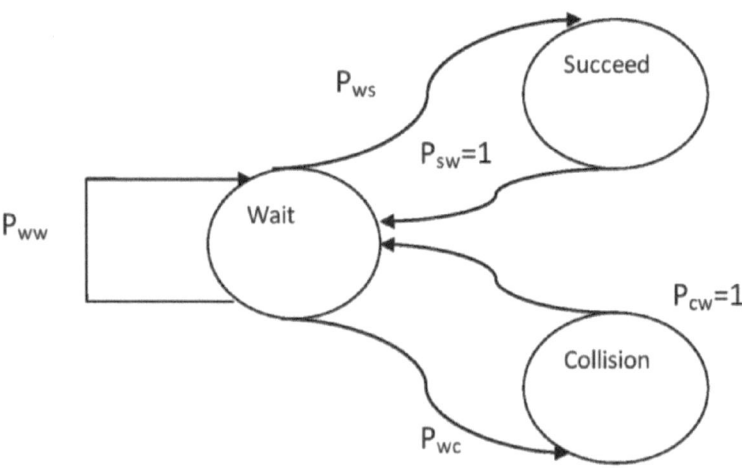

FIGURE 10.3 Markov chain model for the node.

Thus, the translation probability P_{ww} is given by

$$P_{ww} = 1 - p' \tag{10.9}$$

Also, the duration of a node in wait sate T_{wait} is τ (this waiting time is only after the node satisfies the minimum threshold condition). The duration of success and collision states are equal to the frame transmission duration time; hence, T_{succ} and T_{coll} are $\delta_{data} + \tau$. After executing the desired action in success and collision states, node 'x' always returns to the wait sate. Therefore, P_{sw} and P_{cw} equals to 1.

Let Φ_w, Φ_s, and Φ_c represent the steady-state probabilities of wait, success, and collision states, respectively.

$$\Phi_w = \Phi_w P_{ww} + \Phi_s P_{sw} + \Phi_c P_{cw} = \Phi_w P_{ww} + 1 - \Phi_w \tag{10.10}$$

Hence,

$$\Phi_w = \frac{1}{2 - P_{ww}} \tag{10.11}$$

Based on the above condition, transition probability,

$$P_{ws} = P_1 P_2 P_3 \tag{10.12}$$

Where,

$P_1 =$ Prob (node x transmits in a slot)
$P_2 =$ Prob (all of node x's neighbor nodes do not transmit the same slot)
$P_3 =$ Prob (nodes in potential hidden area do not transmit for $2\delta_{data} + \tau$)

Last term represents the vulnerable period that is equal to $2\delta_{data} + \tau$ and $\delta_{data} + \tau$ in the case of slotted ALOHA-based MAC protocols. Obviously, $P_1 = p'$ and P_2 is given by,

$$P_2 = \sum_{i=0}^{\infty} (1 - p')^i \frac{(\lambda \pi R^2)^i}{i!} e^{-\lambda \pi R^2} = e^{-p' \lambda \pi R^2} = e^{-p'N} \tag{10.13}$$

Let A_{tx}, A_{cs}, and A_{ph} represent the transmission area, additional carrier sensing area and potential hidden node area, respectively. As shown in Figure 10.4, additional carrier sensing area is the physical carrier sensing area which is larger than the transmission range and smaller than potential hidden node area.

The potential hidden node area,

$$A_{ph} = 2\pi(2R)^2 - A_{tx} - A_{cs} = 2\pi(2R)^2 - \pi R^2 - A_{cs} = 3\pi R^2 - A_{cs}. \tag{10.14}$$

Hence,

$$0 \le A_{ph} \le 3\pi R^2. \tag{10.15}$$

Let N_{ph} represents the number of node in the potential node area.

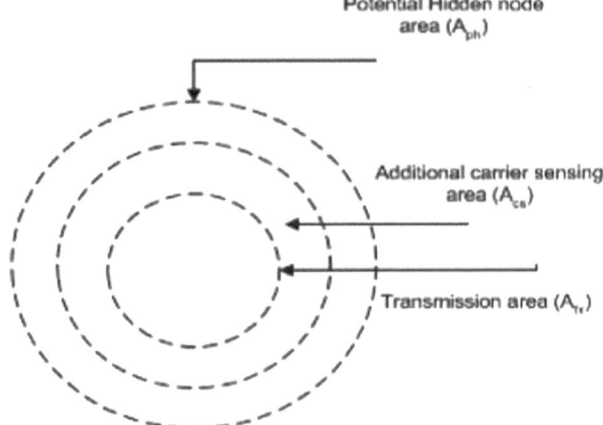

FIGURE 10.4 Transmission area, additional carrier sensing area, and potential hidden node area.

Since nodes are uniformly distributed, hence,

$$N_{ph} = \lambda A_{ph}, \quad 0 \le N_{ph} \le \lambda 3\pi R^2, \quad 0 \le N_{ph} \le 3N \tag{10.16}$$

Hence,

$$P_3 = \left\{ \sum_{i=0}^{\infty} (1-p)^i \frac{(N_{ph})^i}{i!} e^{-N_{ph}} \right\} 1^{(2\delta_{data}+\tau)} = e^{-p'N_{ph}(2\delta_{data}+\tau)} \tag{10.17}$$

Therefore,

$$P_{ws} = p'e^{-p'N} e^{-p'3N_{ph}(2\delta_{data}+\tau)}. \tag{10.18}$$

Since,

$$P_{ws} = 1 - P_{ww} - P_{wc} \quad \text{and} \quad P_{cw} = P_{sw} = 1 \tag{10.19}$$

Hence, the steady-state probability of the success state,

$$\Phi_s = \Phi_w P_{ws} = \frac{P_{ws}}{1+p'} \tag{10.20}$$

Since throughput is the fraction of time in which the channel is engaged in successful transmission of user data. Therefore, the throughput *Th* is equal to the limiting probability that the channel is in the success state.

$$Th = \frac{\Phi_s \delta_{data}}{\Phi_s T_{succ} + \Phi_c T_{coll} + \Phi_w T_{wait}} \tag{10.21}$$

By substituting all the values, hence,

$$Th = \frac{P_{ws}\delta_{data}}{p'T_{coll} + T_{wait}} = \frac{\left(p'e^{-p'\left(N+3N_{ph}\left(2\delta_{data}+\tau\right)\right)}\right)\delta_{data}}{\tau + p'\left(\delta_{data} + \tau\right)} \qquad (10.22)$$

Therefore, the performance of the broadcast scheme is analyzed by varying the average number of neighboring node (N) and transmission attempt probability (p'). In this analysis, the data frame is fixed as 100τ.

Figure 10.5 shows the throughput results of the IEEE 802.11 broadcast scheme with different potential hidden node area R_{ph}

From Figure 10.6, it is clear that as the percentage of A_{ph} reduces to 0, throughput capacity performance increases to the maximum value. This means that by achieving the maximum value of A_{cs}, the IEEE 802.11 broadcast scheme minimizes the probability of hidden node problem. It shows the throughput capacity results of the threshold condition-based broadcast scheme with different potential hidden node area. These threshold conditions help improve the throughput by not letting all the nodes to transmit in the same slot at the same time. With these threshold conditions, it achieves nearly twice of the throughput compared to IEEE 802.11 broadcast scheme.

Figure 10.7 shows the throughput capacity results of the slotted ALOHA-based broadcast scheme. Slotted ALOHA-based broadcast scheme with threshold conditions give quite good throughput capacity.

Figure 10.8 shows the combined results of all the schemes with all the variations in hidden node area. Hence, it is clear that the threshold condition-based broadcast scheme with 0% hidden area gives the highest throughput capacity. So it is beneficial to set the large carrier sensing range for broadcast communication.

10.2.2 MULTI-HOP SCHEME

By virtue of the strategy employed for reducing the collision probability of the packets transmitted from the stations attempting to access the channel simultaneously, a random process given station $b(t)$ is used to represent the backoff counter and is decremented at the start of every idle backoff slot and when it reaches zero, the station transmits and a new value for $b(t)$ is set. The value of $b(t)$ after each transmission depends on the size of the contention window from which it is drawn. Therefore, it depends on the stations' transmission history, rendering it a non-Markovian process. To overcome this problem and to get the definition of a Markovian process, a second process $s(t)$ is defined representing the size of the contention window from which $b(t)$ is drawn [$W_i = 2^i W$, $i = s(t)$]. A backoff time counter is initialized depending on the number of failed transmissions for the transmitted packet.

It is chosen in the range [0, $W_i - 1$] following a uniform distribution, where W_i is the contention window at the backoff stage i. At the first transmission attempt (i.e., for $i = 0$), the contention window size is set equal to a minimum value $W_0 = W$, and the process $s(t)$ takes on the value $s(t) = i = 0$. The backoff stage i is incremented in unitary steps after each unsuccessful transmission up to the maximum value m, while the contention window is doubled at each stage up to the maximum value $W_{max} = 2^m W$.

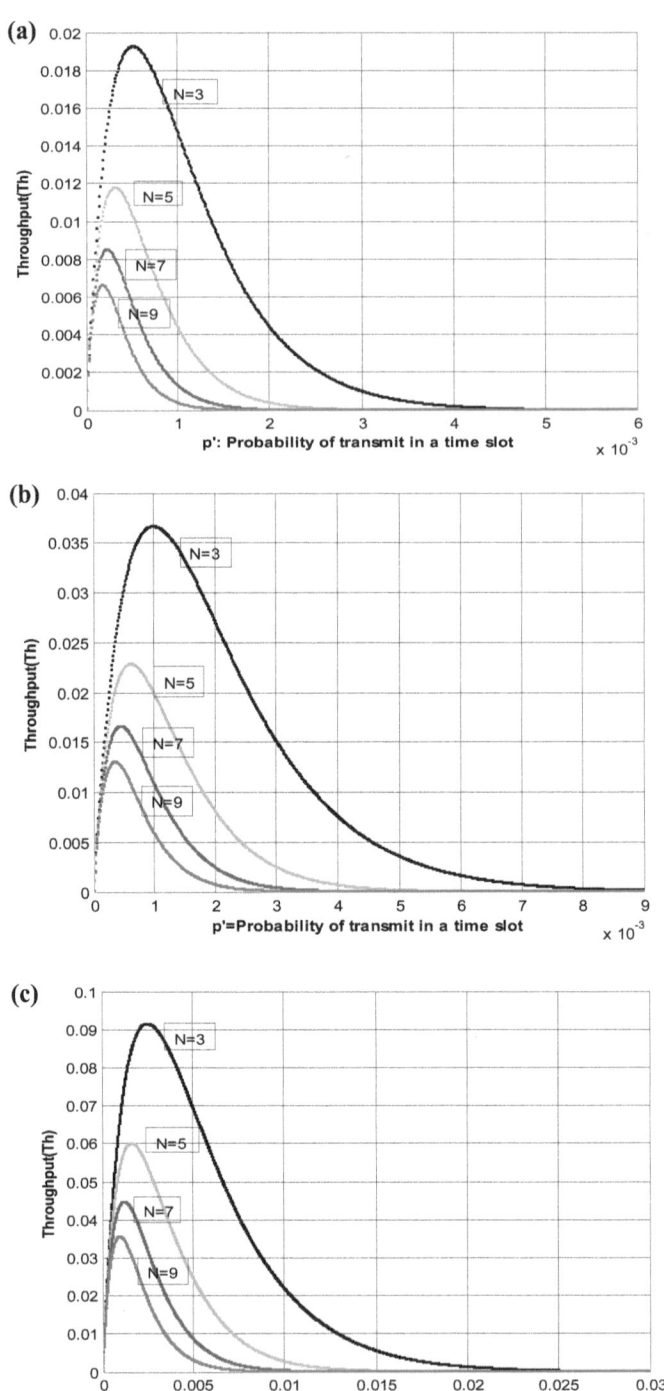

FIGURE 10.5   IEEE 802.11-based broadcast scheme.

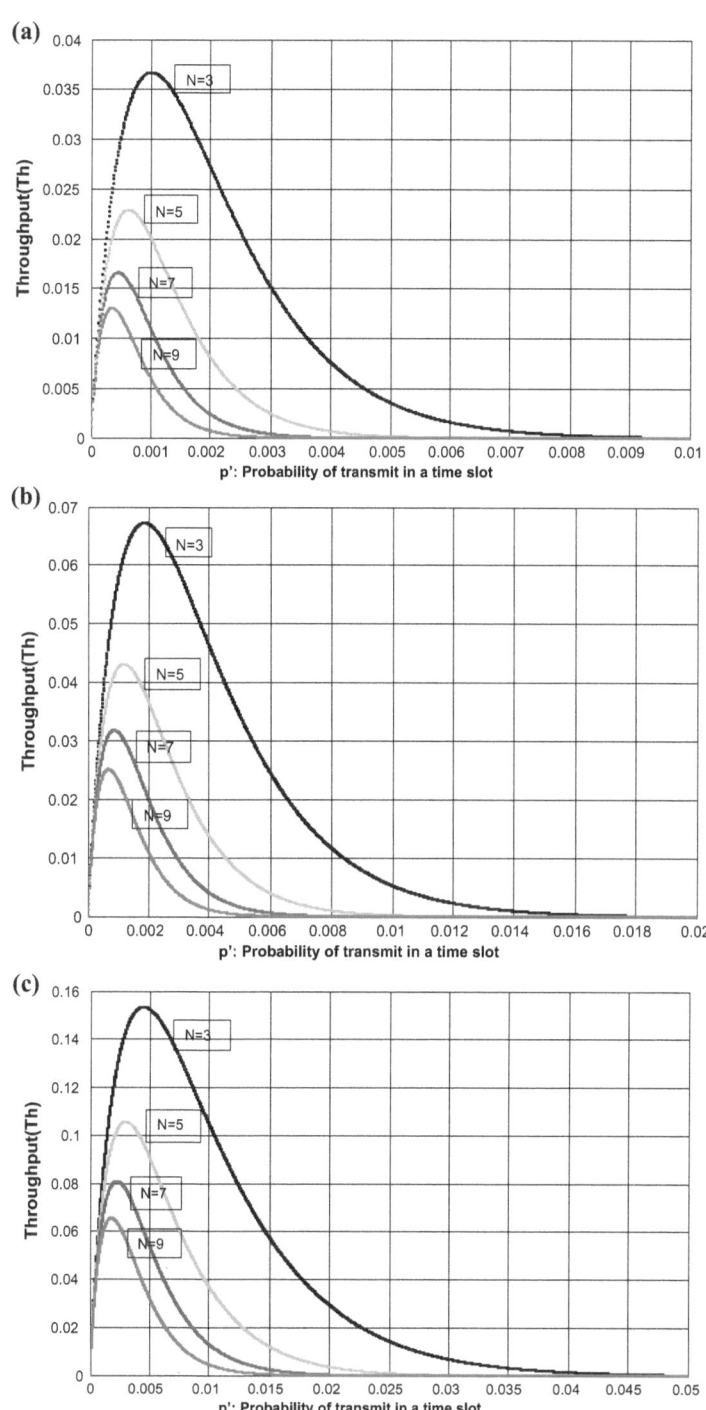

FIGURE 10.6 Threshold condition-based broadcast scheme.

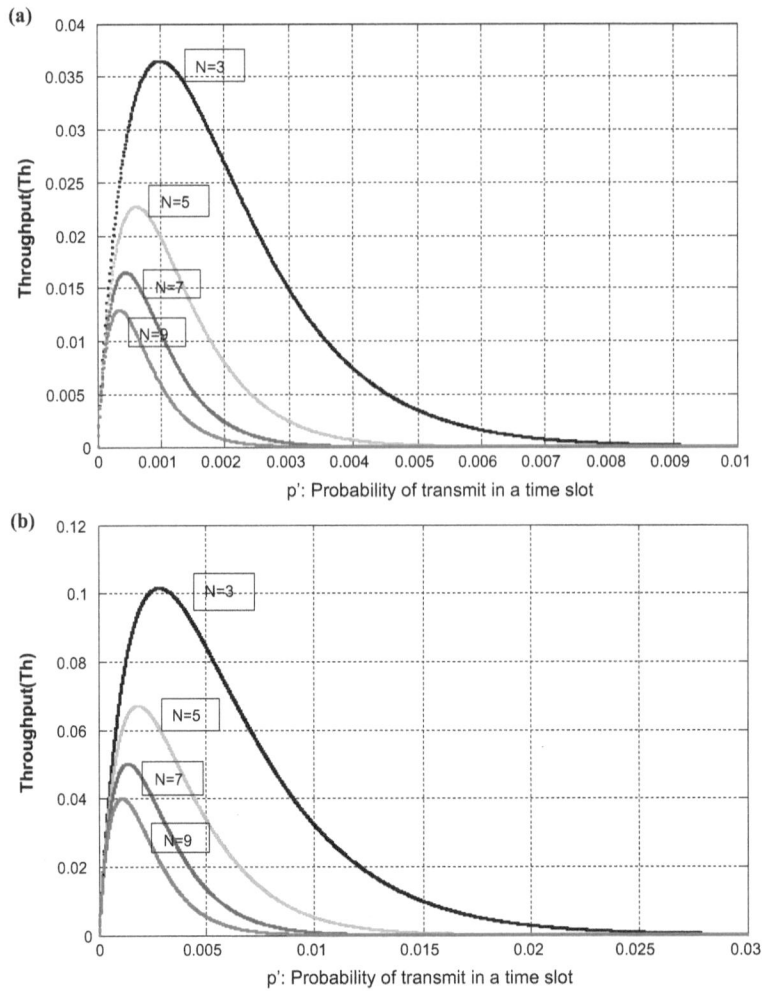

FIGURE 10.7   Slotted ALOHA-type broadcasting scheme.

The backoff time counter is decremented as long as the channel is sensed idle and stopped when a transmission is detected. The station transmits when the backoff time counter reaches zero. A two-dimensional Markov process [$s(t)$, $b(t)$] can now be defined, based on two assertions: the probability τ that a station will attempt transmission in a generic time slot is constant across all time slots; the probability P_{col} that any transmission experiences a collision is constant and independent of the number of collisions already suffered. Collisions can occur with probability P_{col} on the transmitted packets, while transmission errors due to the channel can occur with probability P_e. It is assumed that collisions and transmission error events are statistically independent. In this scenario, a packet is successfully transmitted if there is no collision (this event has probability of P_{col}) and the packet encounters no channel errors during transmission (this event has the probability of $1-P_e$). The probability of successful

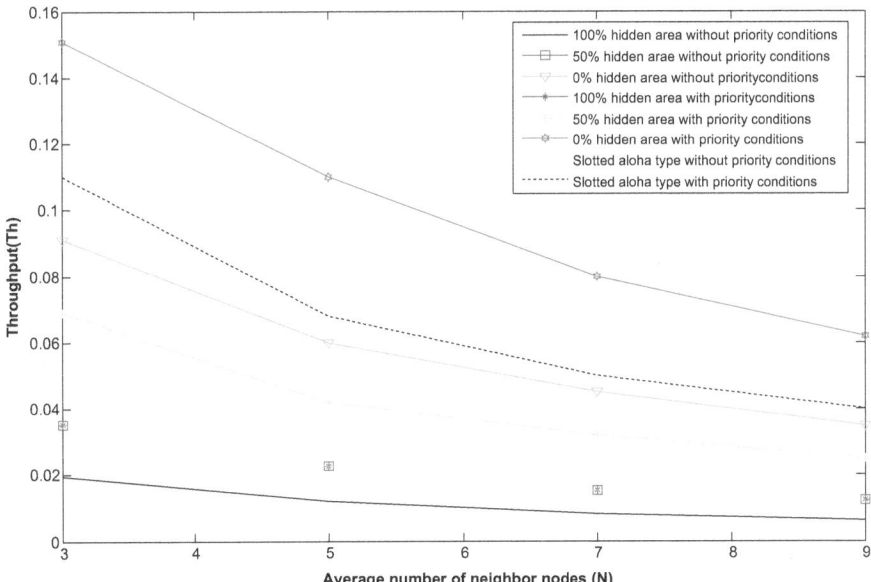

FIGURE 10.8   Combined results.

transmission is therefore equal to $(1-P_e)(1-P_{col})$, from which it can be set an equivalent probability of failed transmission as $P_{eq}=P_e+P_{col}-P_eP_{col}$. The maximum contention windows *(CW)* size is $CW_{max}=2^mW$, and the notation $W_i=2^iW$ is used to define the ith contention window size. A packet transmission is attempted only in the *(i, 0)* states, $i=0, \ldots, m$. If collision occurs or transmission is unsuccessful due to *A* packet transmission, it is attempted only in the *(i, 0)* states, $i=0, \ldots, m$. If collision occurs or transmission is unsuccessful due to channel errors, the backoff stage is incremented, so that the new state can be $(i+1, k)$ with probability P_{eq}/W_{i+1}; since a uniform channel errors, the backoff stage is incremented, so that the new state can be $(i+1, k)$ with probability P_{eq}/W_{i+1} since a uniform distribution between the states in the same backoff stage is assumed. It is considered to be captured as a subset of the event of a collision (i.e., capture implicitly implies a collision). In other word, the capture event can happen in the presence of collision by allowing the transmitting station with the highest received power level at the access point to capture the channel. In the event of capture, no collision is detected and the Markov model transits into one of the transmitting states *(i, 0)* depending on the current contention stage. If no collision occurs (or is detected), a data frame can be transmitted and the transmitting station enters state *(i, 0)* based on its backoff stage. From state *(i, 0)*, the transmitting station re-enters the initial backoff stage $i=0$ if the transmission is successful and at least one packet is present in the buffer, otherwise the station transit in the state labeled *I* waiting for a new packet arrival, as shown in Figure 10.9. Otherwise, if errors occur during transmission, the ACK packet is not sent, an ACK timeout occurs, and the backoff stage is changed to $(i+1, k)$ with probability P_{eq}/W_{i+1}.

To make the system model, following assumptions have been made. The collision probability needed to compute τ (transmission probability) can be found considering

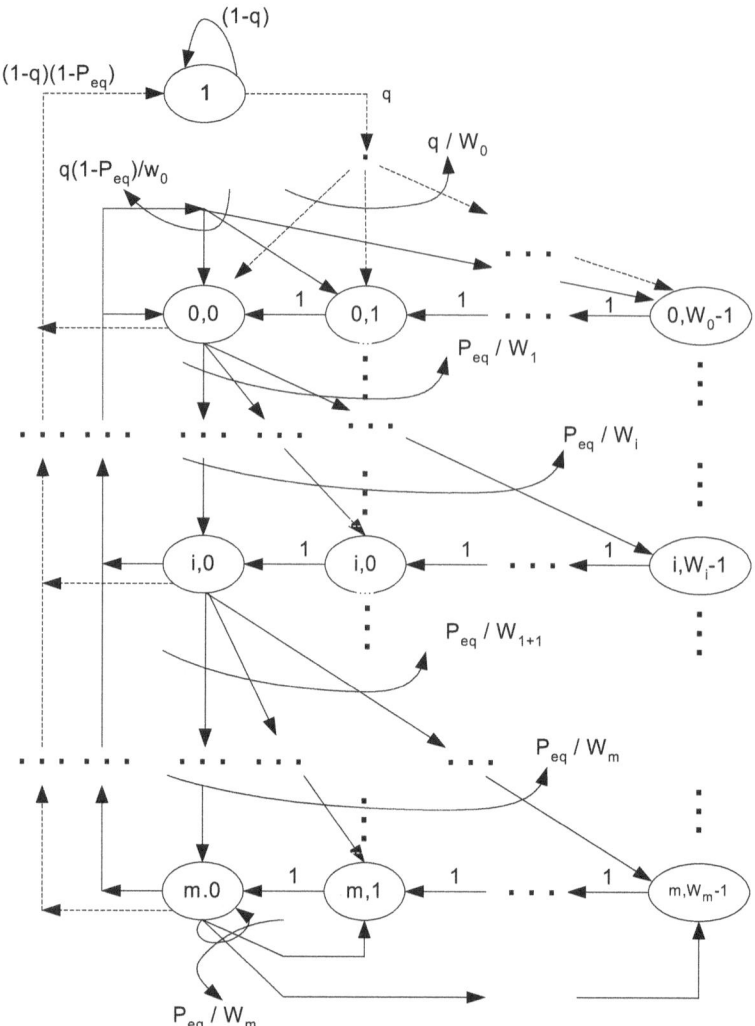

FIGURE 10.9 Markov chain for the contention model.

the aforementioned network topology. A packet from a transmitting node encounters a collision if in a given time slot, at least one of the remaining *(N−1)* node transmits simultaneously another packet, and there is no capture effect.

$$P_{col} = 1 - (1 - \tau)^{N} - 1 \qquad (10.23)$$

It is considered to be a scenario in which N nodes are uniformly distributed in a circular area of radius *R* transmit toward a common base node in the center of such an area.

It computes to normalized channel capacity, defined as fraction of time the channel is used to successfully transmit payload bits:

$$S = \frac{P_t P_s E[PL]}{(1-P_t)\sigma} + P_t(1-P_s)T_c + P_t P_s T_c \tag{10.24}$$

where P_t is the probability that there is at least one transmission in the considered time slot, with N stations contending for the channel, each transmitting with probability,

$$P_t = 1-(1-\tau)^N \tag{10.25}$$

P_s is the conditional probability that a packet transmission occurring on the channel is successful. This event corresponds to the case in which exactly one station transmits in a given time slot. This condition yields the following probability:

$$P_s = N\tau(1-\tau)^N - 1/P_t \tag{10.26}$$

T_c and T_s are the average times a channel is sensed busy due to a jamming data frame transmission time and successful data frame transmission times, respectively. Knowing the time durations for ACK frames, ACK timeout, σ, data packet length (PL) and PHY and MAC headers duration (H) and propagation delay, τ_p, T_c and T_s can be computed as follows:

$$T_c = H + PL + \tau_p \tag{10.27}$$

$$T_s = H + PL + \tau_p + ACK \tag{10.28}$$

It assumes that packet will be dropped after few retries. It is also assuming packet sequence, as shown in Figures 10.10 and 10.11.

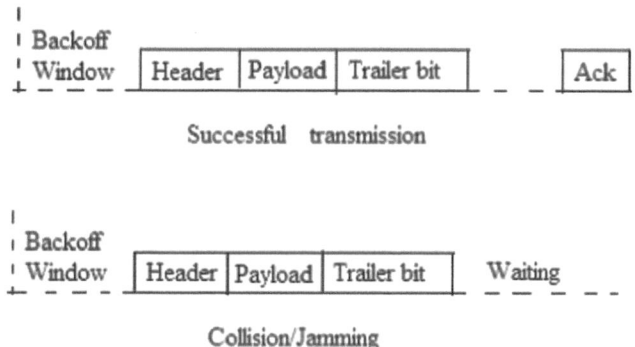

FIGURE 10.10 Packet timing diagram.

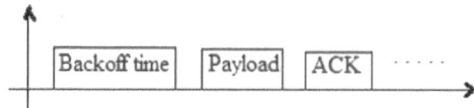

FIGURE 10.11 Packet sequence for regular node.

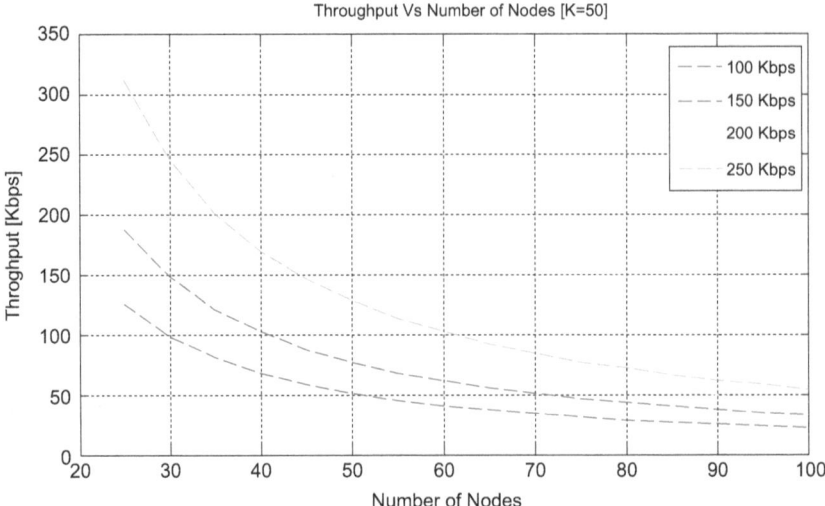

FIGURE 10.12   Lower limit of throughput.

The throughput capacity of ad-hoc wireless network is defined as the maximum number of bits per second that can be reliably transmitted over the channel. Throughput capacity c not only depends on physical conditions such as fading, path loss, topology and property of antennas but also the chosen routing and/or MAC scheme. Hence, the throughput capacity of ad-hoc wireless networks can be expressed as,

$$c = k\frac{T}{n \, \log n} \tag{10.29}$$

where C is the channel capacity, k represents the channel constant and as scaling factor, T represents the available bandwidth/data rate and n is the number of nodes in a network.

Based on numerical model and Equation 10.29, it is calculated as the fundamental lower limit of throughput capacity, as shown in Figure 10.12. The graph shows the throughput capacity performance for different values of data rate and k as the number of nodes is increasing the throughput start decreasing. It has been observed that nature of graph remains same after varying different parameters and throughput starts decreasing as number of nodes are increasing.

10.3   SIMULATION MODEL

In this simulation model, the traffic model is taken as the constant bit rate (CBR) as in most of the application, CBR is widely used traffic, for example, text, browsing, etc. Variable bit rate (VBR) can be considered for video communication if needed. At MAC Layer, the assumed bandwidth of 20 MHz is at 2.4 GHz band (open/unlicensed band). Power has been considered as 53 dBm (4 W) (FCC Standard) and in India, approximately three times (i.e., 6–7 W), mobile: 0.1 mW. To improve the better performance of the AODV routing protocol, the nodes are being increased up to 100. Initially for this number of nodes, the simulation is running for ten transmissions till it achieves statistical stability up to 95%. Therefore, the performance of existing AODV and modified AODV (AODV+) simulation results has been compared at different simulation times, i.e., 150 and 250 s (Tables 10.1 and 10.2).

By referring the observation table and graph of average throughput for simulation times of 150 and 250 s, respectively, it is observed that as the nodes are increasing, the performance of AODV routing protocol is fluctuating, i.e., it may remain same for some values of nodes and then is suddenly increasing or decreasing for some certain values of nodes, whereas the performance of AODV+ routing protocol provides better results as the number of nodes are increasing, as shown in Figure 10.13. Therefore, AODV+ is more stable due to back up routes for packets since it is showing expected results as an on-demand protocol.

TABLE 10.1
Parameters Assumption (up to 100 Nodes)

Nodes	AODV (150 s)	AODV+(150 s)	AODV (250 s)	AODV+(250 s)
25	667.36	537.19	252.19	389.11
40	668.48	615.38	628.26	786.14
55	391.85	480.71	667.47	759.24
70	503.79	689.21	667.96	729.13
85	274.40	478.35	667.65	754.31
100	335.17	450.15	528.75	677.34

TABLE 10.2
Observation Table of Average Throughput (up to 100 Nodes)

Sr. No.	Parameters	Values
1	Traffic model	CBR (constant bit rate)
2	Area of experiment	500 × 500 m
3	Number of nodes	25 ~100
4	Routing protocol	AODV and DSR
5	Packet size	1,000 bytes
6	Simulation time	150, 250 s
7	Mobility scenario	5 km/h
8	Mobility model	Random way point
9	MAC layer protocol	IEEE 802.11 with 2.4 GHz

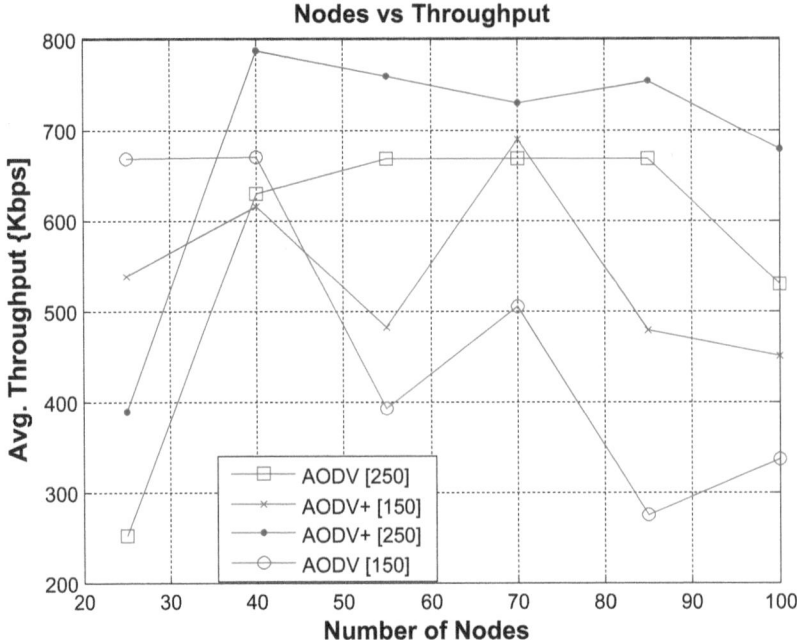

FIGURE 10.13   Average throughput.

10.4   RESULT

Figure 10.14 shows the graph of number of nodes versus throughput for AODV, AODV+ and lower limit. It has been observed that AODV+ is giving better performance for throughput because AODV+ has created routing table considering link quality for the same. Furthermore, the results have identified the optimization of the network performance with 50 nodes. As shown in Figure 10.15, AODV+ has gained 108 times throughput compared to the lower limit and AODV gained 32 times of the lower limit. So overall there is 54 times better throughput gained of channel capacity using AODV+.

10.5   CONCLUSIONS

To satisfy the demand of high data rates with limited resources is a big challenge. One solution to this challenge is to optimize the use of available resources and hence it is required to enhance the performance of networks. This research has multiple objectives.

Therefore, the research contribution of this work has been mentioned in the following ways:

- It has been surveyed in a detailed way and has discussed about their proposals of limitations which do not exist in the current literature survey as per the best of our knowledge.

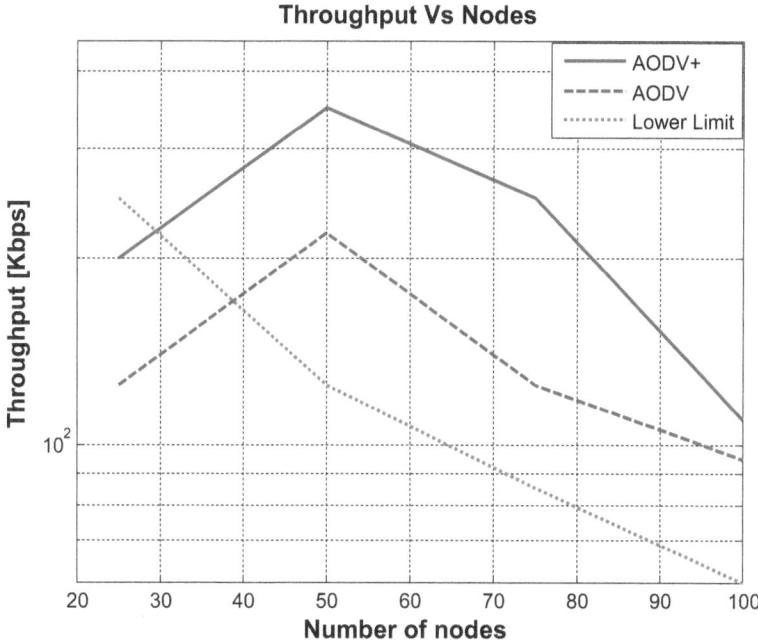

FIGURE 10.14 Analytical and simulation comparison of average throughput.

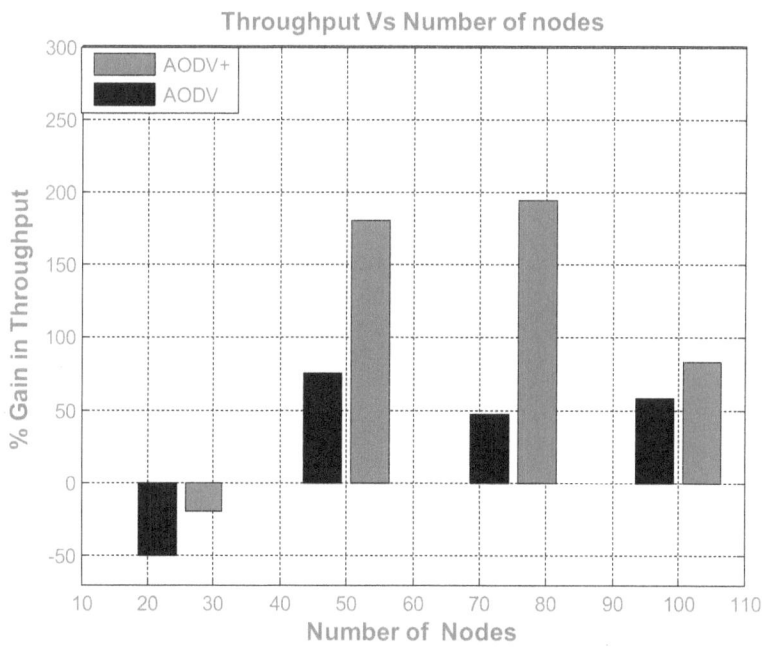

FIGURE 10.15 Comparison of AODV and AODV+ of average throughput.

- It has carried out extensive simulation survey to compare existing routing protocols for the detailed study.
- It has developed an analytical model to give theoretical bounds of average throughput to get a more practical approach.
- It has proposed a modified AODV (i.e., AODV+) routing protocol to increase the performance of average throughput by verifying on various parameters in ad-hoc wireless networks.
- It has also checked on the proposed idea with the detailed study of routing protocols to see the presence in different conditions.
- It has done the performance analysis of the proposed idea with mobility and without mobility.
- It has compared the large scalability till nodes of 100 which has been addressed in very few research articles to the best of our knowledge.

REFERENCES

1. C-K. Toh, *Ad Hoc Mobile Wireless Networks: Protocols and Systems*, Pearson Education, 2013.
2. D. Dhokal, K. Gautam, "Performance Comparison of AODV and DSR Routing Protocols in Mobile Ad-hoc Networks: A Survey", *International Journal of Engineering, Science and Innovative Technology (IJESIT)*, vol. 2, No. 3, 258–265, 2013.
3. H. Bakht, "Survey of Routing Protocols for Mobile Ad-hoc Network", *International Journal of Information and Communication Technology Research*, 258–270, 2011.
4. R.E. Rezagah, A. Mehmood, *Analyzing the Capacity of Wireless Ad-hoc Networks*, IEEE Transaction, 2009.
5. S. Mehta, J. Kim, "Improved Sensor MAC Protocol for Sensor Networks", *International Journal of Lateral Computing (IJLC)*, published by MacMillan India ltd. ISBN-10:1403-9310-1, 2006.
6. J.M. Choi, J. So, Y.B. Ko, "Numerical Analysis of IEEE 802.11 Broadcast Scheme in Multihop Wireless Ad-hoc Networks", *ICOIN*, 1–10, 2005.
7. J. Li, C. Blake, Douglas S.J. De Couto, Hu Imm Lee, Robert Morris, M. Grossglauser and D. N. C. Tse, "Mobility increases the capacity of ad hoc wireless networks," in IEEE/ACM Transactions on Networking, vol. 10, no. 4, pp. 477–486, Aug. 2002, *IEEE*, 2002.
8. P. Gupta, P.R. Kumar, "The Capacity of Wireless Networks", *IEEE Transaction on Information Theory*, vol. 46, No. 2, 388–404, 2000.
9. L. Wu, P.K. Varshney, "Performance Analysis of CSMA and BTMA Protocols in, Multihop Networks(1). Single Channel Case", *Elsevier Information Science*, vol. 120, 159–177, 1999.
10. R. Rom, M. Sidi, *Multiple Access Protocols: Performance and Analysis*, Springer-Verlag, 1989.

11 Advanced Metering in Smart Distribution Grid

Gaurav Srivastava
Poornima College of Engineering

CONTENTS

DOI: 10.1201/9781003143802-11

11.1 RATIONALE

An electric power grid is a system of intensity generators, transmission lines, transformers, and appropriation/transfer frameworks to give its purchasers (private, mechanical, and business) with the force they need. As of now, electrical vitality is produced in unified force plants and shipped over a long-distance transmission system to dissemination systems before arriving at the end purchasers by means of correspondence and force streams in just a single way, i.e., from power plants to the clients, which is on the whole called an electric framework. After numerous times of advancement, it has been understood that different utilities can interconnect to accomplish more prominent dependability of by and large force framework by making up for surprising disappointments just as separations from power gadgets, i.e., transmission lines and generators.

In an electric framework, age, transmission, and appropriation of intensity ought to be unequivocally planned. Figure 11.1 portrays different segments in the present electric framework, which comprises four regions that are age, transmission, conveyance, and customers. Distribution incorporates taking the power from transmission lines and conveying it to the clients. Commonly, a power conveyance framework incorporates medium voltage electrical cables (under 50 kV), substations, and transformers, beginning at the transmission substations and completion at the meters of clients.

Besides, the worldwide environmental change and ozone harming substance outflows on the Earth brought about by the power and transportation ventures put more weight on the current force frameworks. Therefore, another idea of next-generation electric force framework is direly expected to address these difficulties, which spurs the proposition of smart grid (SG).

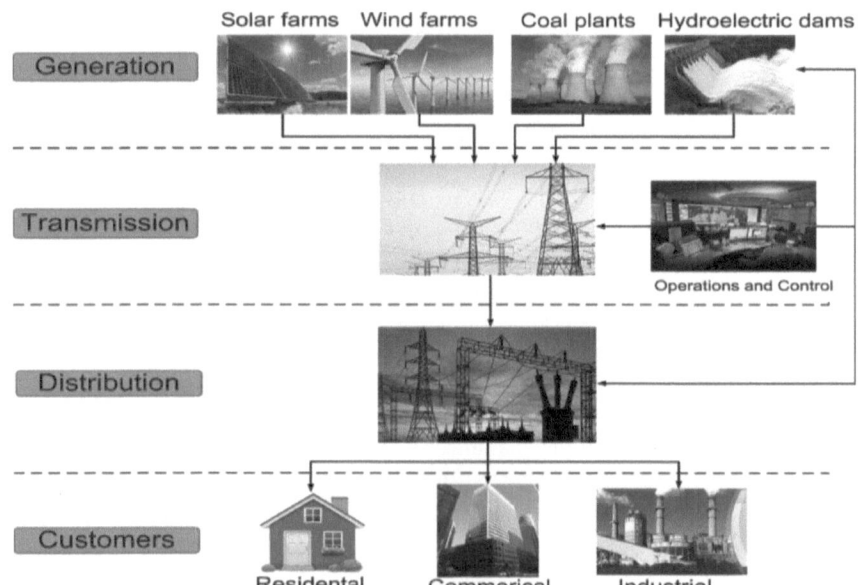

FIGURE 11.1 Typical electric power grid.

The smart grid can be seen as a superposition of correspondence systems on the electric networks. Consequently, it can improve proficiency, dependability, wellbeing, and security of power gracefully to the clients, with a consistent coordination of inexhaustible and elective vitality sources, for example, photovoltaic frameworks, wind vitality, biomass power age, flowing force, little hydropower plants, and plug in half and half electric vehicles, through computerized control and present day interchanges advancements

11.2 OBJECTIVE OF THE PROJECT

- This venture will create test methods and conventions that apply genuine waveforms that are powerfully changing and have fluctuating measures of twisting present alongside bidirectional force streams to shrewd meters. The task will likewise lead tests to guarantee that meter exactness isn't influenced by impedance from the correspondence module in the meter.
- To build up a savvy meter testing office to create execution test necessities to guarantee keen meters to be precise under field conditions.
- To propel estimation science for the presentation of savvy meters and related metering guidelines and test systems that will empower exact and reasonable power metering required for the keen lattice.

11.3 LITERATURE REVIEW

This chapter provides some brief ideas about the different profound studies that had been performed in various fields.

The study of these papers helps us in analyzing the previous studies, improvements in booster design, and their applications in various fields like WECS, PV arrays, its use in electric vehicles etc., and election of appropriate components of better results.

The papers that have been reviewed basically focus on the actual model and other sub papers which are related to the different components used in the model. The papers cover the range from year 2009 to 2020 and are mostly between 2014 and 2020. The papers used are from different conferences including IEEE, ScienceDirect, IJERT, GRO, etc.

In this chapter, further we are going to discuss the basic problems which had been discussed in various papers, methodologies which had been adopted to solve the problems and their future scopes.

11.4 REVIEW PROCESS ADOPTED

The data that the literature reviewer gathers to frame a writing audit is named as information. Hence, the writing audit procedure can be viewed as an information assortment process, i.e., as a methods for gathering a group of data with respect to a specific subject of intrigue. As an information assortment device, the writing audit includes exercises, for example, getting, recording, distinguishing, and sending data. To be sure, the writing audit process is realized through information assortment.

In the field of research, the term method refers to the specific approaches and procedures that the researcher systematically utilizes in the research design, sampling design, data collection, data analysis, data interpretation, and so forth.

11.5 METHODOLOGY

Advanced metering infrastructure also called AMI is the aggregate term to portray the entire framework starting from the smart meters to the two-way communication system to control focus equipment and all the other applications that empower the accumulation and transmission of energy utilization data in coming real time. It forms two-way communication with clients conceivable and is the foundation of smart grid. The destinations of AMI can be far off meter perusing for mistake-free information, arrange issue recognizable proof, load profiling, vitality review, and incomplete burden reduction instead of burden shedding. The AMI incorporates shrewd meters, e.g., electric, gas, and warmth meters, at client premises, passages, correspondence spine arrange among client and specialist organizations, and information of the board frameworks to quantify, gather, oversee, and break down the information for additional preparing. Smart meters can recognize power utilization considerably in more detail than a regular meter and intermittently send the gathered data back to the service organization for load checking and charging purposes. Moreover, the information from savvy meter readings is additionally basic for the control community to execute Demand/Response instrument (Figure 11.2).

AMI is an arranged foundation that incorporates various innovations to accomplish its objectives. The foundation incorporates smart meters and communication systems in various degrees of the framework hierarchy, and Meter Data Management Systems (MDMS) intends to coordinate the gathered information into programming application stages and interfaces.

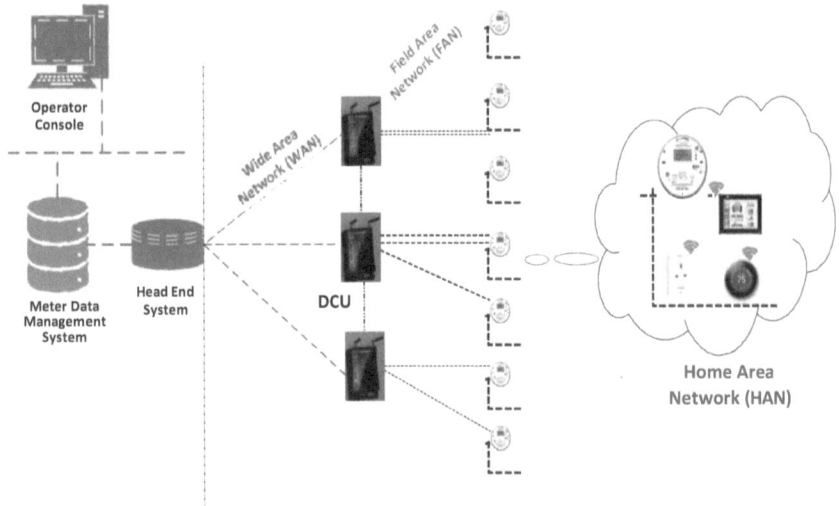

FIGURE 11.2   Advanced metering infrastructure.

AMI in the distribution network can be viewed as comparable to the SCADA/ energy management system (EMS) that manages the transmission network. At distribution part of grid, AMI works as a smart grid technology. Information is sent in the real time at the utility-controlled centers because advanced metering incorporates a communication layer in the ongoing distribution grid. As opposed to the SCADA information which checks the information pretty much every 2 s, AMI framework gathers information at a time period minutes. This constant 15-min information can be utilized in numerous applications like force quality observing, deficiency area identification, blackout the board frameworks, and so forth. AMI framework can relieve a significant number of the issues identified in the appropriation framework. Investigation of the ongoing information depicts the main reasons of intensity increase at system. Likewise, the voltages of system of feeders could be promptly acquired out of the brilliant systems of meters constant information. Voltage levels at the distant end of hubs of the feeding part might be escalated by fittingly altering the levels of voltage controllers and burden of the tapping transformers and along these lines lessening AT&C misfortunes. Force burglary in a feeder can be recognized by contrasting the feeder utilization information against the aggregate utilization of hubs under that specific hub.

Architectures of implementing AMI:

1. using some sort of the technology included cellular means, smart meters communicate directly to a head end system
2. using low-power communication technologies, smart meters communicate with a data concentrator unit that transmits the information to the utility servers via cellular networks (Figure 11.3)

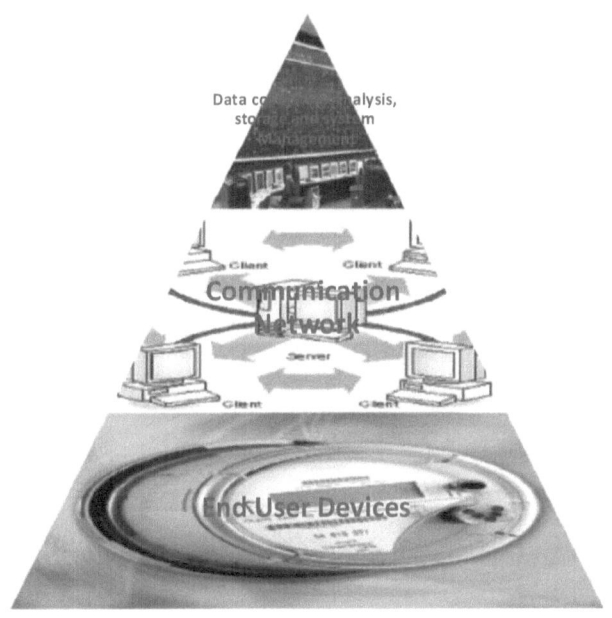

FIGURE 11.3   Schematic representation of AMI.

11.6 AMI ARCHITECTURE

Smart meters are a propelled energy meter, that bolsters two-way interchanges contrast, and a regular energy meter. Henceforth, it can quantify the vitality utilization information of a purchaser and afterward communicates added data to the service organizations to help decentralized age sources and vitality stockpiling gadgets and bill the client in like manner. In addition, shrewd meters can get data about power cost and orders from service organizations and afterward convey them to shoppers. Practically speaking, brilliant meters can peruse vitality utilization data of clients continuously, for example, estimations of voltage, recurrence, and stage point, and afterward they safely impart the data to control focuses. By utilizing bidirectional correspondence of information, savvy meters can gather data in regards to the power utilization estimations of client premises. Information gathered by keen meters is a blend of boundaries, for example, an extraordinary meter identifier, timestamp of the information, and the power utilization esteems. In view of the data, brilliant meters can screen and execute control orders for every single home gadget and apparatuses at the client's premises distantly just as locally. In addition, shrewd meters can speak with different meters on their arrive at utilizing home territory organize (HAN) to gather symptomatic data about apparatuses at the client just as the dispersion network. Also, keen meters can be modified with the end goal that solitary force devoured from the utility network is charged though the force expended from the dispersed age sources or capacity gadgets claimed by the clients isn't charged. Accordingly, they can constrain the most extreme power utilization and can end or reconnect power gracefully to any client distantly (Figure 11.4).

Main components of AMI are:

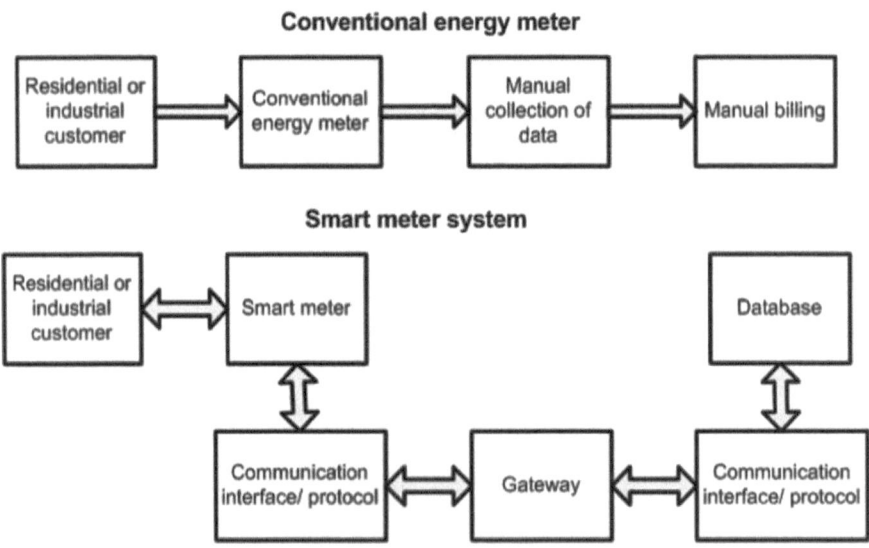

FIGURE 11.4 Architectural model of conventional energy meter and smart meter.

11.6.1 Smart Meters

Smart meters are fundamental building block of AMI. It may be an electric meter that is fitted with an outside modular part or very well may be independent with all the metering, control and communication modules embedded in a solitary unit. The guidelines by Cea command the choice utilized for actualizing. A controlled relay is given to distant association/detachment. This far off associate/detach utilized executing usefulness and disengaging on detection. They additionally permit expanded goal of information on different estimation boundaries over the framework and this information utilized by the accompanying uses:

- Blackout identification, reaction, reclamation giving information to the field tasks opportune
- Keeping clients better educated about the status of the intensity matrix. Utilities can impart significant data, e.g., reason for blackout, field-evaluated reclamation time, and open security notice
- Improving versatility against interruptions, lessening expected blackouts, decreasing recurrence and span of blackouts by improving exactness of the matrix resource arranging and the board.

11.6.2 Data Concentrator Unit – (DCU)

It collects information from a group and sends it to HES. It works as a doorway among those meters that are downstream and the upstream HES. Also, encourages bidirectional communication information to the head framework and thus sharing controlling instructions from the HES to the smart meters.

11.6.3 Meter Data Acquisition System – (MDAS) or Head End System – (HES)

Server-based meter information head end framework is in agreement with different standard-based conventions just as restrictive conventions. MDAS trade meter information to the board frameworks combined with the examination on the standard information trade model.

The fundamental goal of HES is to gain meter information automatically maintaining a strategic distance from any human intercession and screen boundaries obtained. Pulls out the information from the savvy meter at standard spans, and furthermore on-request. Goes about as impermanent crude information can be put away for a span b information on the board framework. Monitors specialized gadgets and updates the system geography data. It likewise makes retry endeavors to set up correspondence with the DCU in the event that there is a correspondence interface disappointment.

11.6.4 Meter Data Management System (MDMS)

The framework involves gathering information at that point preparing and putting away them appropriately. Thus, as result, same kind of information are obtained from a similar location and it gets simpler to recover the necessary information. MDMS incorporates scientific apparatuses which empower diverse segment of activity and the executive's framework to collaborate with it and gather whichever information is required. Activity and the board framework mostly incorporate: Outage of the executive's framework (OMS), Consumer data framework (CIS), Geographic data framework (GIS), and Distribution of the executive's framework (DMS), AMI correspondence is situated focal activity community. Since it brought together, from a solitary worker activity, the executive framework prepares items. The center of AMI is a meter information the executives framework exhibit. Regularly, the information is put away for utilization in investigation in different areas. Primary elements are information assortment, information stockpiling, information trade, approval, altering and estimation of information gathered, charging, report age, alerts/occasions of the executives. The approved information can be utilized in applications and administrations, for example, load investigation/load anticipating, request reaction (top burden the board), blackout the executive's framework, prepaid usefulness, circulation transformer wellbeing checking framework, and so forth.

11.6.5   HOME ENERGY MANAGEMENT SYSTEM – HEMS

Screens and streamlines vitality utilization in a brilliant house. Sews procures the force utilization information from the shrewd fittings introduced between the force gracefully and apparatuses. Additionally, clients can see their vitality bills and complete vitality use continuously since the HEMS can associate and get the information from brilliant meter. Since the clients can follow their utilization, this empowers them to improve vitality use to limit their bills. Truth be told, request reaction, which is one key highlights of the keen lattice is acknowledged utilizing HEMS. Utility may advise for blackouts, lead clients to vitality proficiency giving data over vitality sparing.

11.7   AMI COMMUNICATION TECHNOLOGY

Different geographies and models can be utilized for correspondence in smart grids. The most drilled engineering is to gather the information from gatherings of meters in nearby information concentrators and afterward communicate the information utilizing a backhaul channel to headquarters where the workers were putting away the information and preparing offices just as the board and charging applications dwell.

As various kinds of structures and systems are accessible for the acknowledgment of AMI, there are different mediums and correspondence advancements for this reason too. Some are:

- Power Line Carriers (PLC).
- Broadbands over Power Lines (BPL).
- Copper/optical fibers
- WiMaxs
- Bluetooth
- General Packets Radio Services (GPRS).

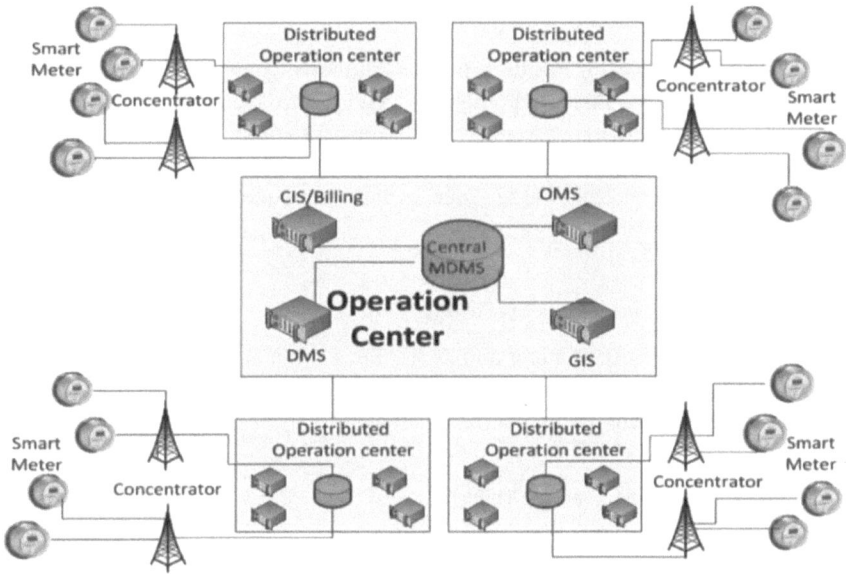

FIGURE 11.5 A distributed communication and management architecture in smart grid.

- Satellites
- Peer-to-Peers
- Zigbees (Figure 11.5)

11.7.1 WIRELINE COMMUNICATION TECHNOLOGY

The information collected is transmitted to HES on the WAN. Because the utility is providing it to a wide area, it can be far from the DCU and may spend many kilometers, therefore long-range communication like the cellular communication, satellite communication, and optical fiber will fit most.

The wireless correspondence systems are one of the most widely investigated themes in power frameworks associated with SG idea. In spite of the fact that the remote systems got a few favorable circumstance terms of establishment and inclusion, the basic lacking is their affectability to restricted transfer speed and obstruction. The remote system is involved in progressive work systems utilizing remote LANs to connect with electrical gadgets. The most appropriate AMI foundations are NANs and HANs for remote arrangement inferable from their minimal effort establishments. The web-based correspondence foundations and information of the executives focus (DMPs) can be introduced either wireline or remote where the correspondence among NANs and DMPs can cover a scope of kilometers. Any DMP can interface and oversee several SMs where a wide inclusion territory can be introduced with work organizing or handing-off the DMPs. The creative investigations in SG applications depend on exceptionally adaptable and organize wide spreading correspondence that can be handily established by utilizing remote sensor systems (WSNs).

11.7.2 Home Area Networks (HAN)

Intended gadgets imparting inside a house. Regularly comprises brilliant fittings, keen indoor regulators, shrewd apparatuses. Savvy introduced the client area likewise. Furthermore, IHD likewise speaks with the brilliant meter for levy and aggregate utilization data. It is through HAN that projects, for example, conduct request reaction, dynamic levies, and burden management for request administration can be executed.

11.7.3 Field Area Networks (FAN)

FAN comprises the brilliant meter and DCU. Goes about entryway and the brilliant meter. It encourages a bidirectionally correspondence information and gets orders through. The mainstream remote correspondence advances utilized.

11.7.4 Wide Area Networks (WAN)

Interfacing Dcu's structure zone organize. Information accumulated is conveyed providing wide geological territory, and community may be extremely distant situated. Along these lines, it may traverse many kilometers and subsequently correspondence alternatives, for example, the cell interchanges, satellite correspondence, and the prerequisites.

11.8 WORKING PRINCIPLE OF SMART METER

Smart meters incorporate a blend of equipment, programming, and alignment frameworks. Correspondence are the center components. Presents a square chart of brilliant meter indicating structure hinders for a total keen metering arrangement. A brilliant meter framework, information correspondence module, power the executive's framework, recognition, and the focal point of a savvy meter equipment depends on the framework design help the estimations. The simple finish comprises simple to computerized converter (ADC) sources of information. An incorporated increase stage gives additions to low-yield sensors. An equipment (be utilized to additionally quicken math-concentrated activities during vitality calculation). Then again, the product bolsters computation of different boundaries, responsive force, and recurrence are the fundamental boundaries determined during the vitality estimations.

1. Analog part of smart meter powering supply:
   permits the activity vitality fueled legitimately. Against associating channel: the simple front of insurance that demonstrates like an enemy of moniker channel. The counter false name sides are extraordinary, in light of the fact that the information to maintain a strategic distance from huge stage move. A resistor, voltage is estimated over that is utilized current estimation. Hostile to altering hardware: is available in the simple piece of keen meter. The most widely recognized procedure is utilizing two current sensors for alter discovery. The simple voltage, current signs are prepared by

the L1 simple to-computerized sound's examination: investigation to lessen misfortune. Three procedures of music end from the simple signs are: utilizing band restricting channel.

2. Digital side of smart meter:

computerized segment of a shrewd meter incorporates and is the core of the vitality, it does the computations of the qualities, communicates information as per guidelines of the information organization, according to solicitation.

11.9   ADVANCED METERING'S SECURITY CHALLENGES

As the quantity of smart meter increases exponentially, security issues related with SG and AMI develop considerably from inside the framework just as outside. Nitty gritty data of clients' utilization is basic as it can uncover their way of life. Transmission of information over significant distance just as putting away the information in different spots for re-transmission or investigation can likewise make weaknesses as far as information burglary or control. The value sign and orders got at shopper end are additionally possible territories for digital and physical assault with the end goal of secret activities, harming foundation or force robbery. Besides, shopper's genuine feelings of serenity are basic in the achievement of shrewd meters and development of AMI. On the off chance that shoppers accept that their own information is utilized without wanting to, or experience helpless assistance or force quality because of outside control of the framework by unapproved gatherings or programmers, at that point they, in all likelihood, oppose the execution of AMI. Potential wellbeing risks and higher bills after establishment of such shrewd meters will likewise influence the shoppers' choice. The legislature is paying attention to these issues and is taking a shot at strategies to ensure clients' security of data. The administration is likewise propelling efforts to build the open's information on shrewd meters and to address their authentic concerns in regards to wellbeing and cost issues. Service organizations just as establishment professionals are additionally assuming a significant job in such a manner.

11.10   SMART METERING FUNCTIONS

11.10.1   UTILITIES REQUIREMENT

The main aim of the utilities should be for the highest quality of smart metering, and full advantage should be generated out of the present technology levels by the regulations.

Some main functionalities of smart meters:

- meter reading remotely.
- load profile information.
- remotely managing of meter
- reduction in demand remotely
- connecting/disconnecting remotely

- supply quality
- demand-response and dynamic costing
- detecting power outages.

11.10.2   DYNAMIC PRICING

It is a method to flatten the peak demands. In this method, consumers are motivated to reduce their energy consumptions at the time peak periods. This decreases the wholesale market's prices in a short-term and also customers can lower their electricity bill by switching to off-peak hours.

Three types of dynamic pricing techniques:

1. Time-of-use pricing – TOU
2. Real-time pricing – RTP
3. Critical peak pricing – CPP.

11.10.3   BI-DIRECTIONALLY COMMUNICATING

In this, energy information is been read and then stored by the smart meters during regular time periods. This information is then transmitted through concentrator to the MDMS when it is requested.

11.10.4   REMOTE SERVICES

A smart meter is a digital device that allows remote controlling by utility companies for various power-related issues, like:

- remote power disconnection/connection to consumers
- detecting and correcting theft
- send signals back to consumers
- detecting power outages.

11.10.5   FAULT (OUTAGE) DETECTION

Smart meters help utilities in identifying when and where the outages took place and also notifies that the power has now been finally restored. For this purpose, smart meters need to have these features:

- differentiating capabilities among momentary and sustained parts of event
- verifying capabilities and checking of power statuses
- sending notification about positive power restoration.

11.11   CHALLENGES IN ADVANCED METERING INFRASTRUCTURE

In this particular section, the several challenges of AMIs are recognized. These are divided in two domains, namely, security issues and communication.

11.11.1 SECURITY CHALLENGES

- **Difficulties in recognizing large-scale failures**: arises from the high-level dependence among the components of grid. Very high complexities enlarge the probabilities of flaws. The possibility of attacks resulting in failure, for example, in adversary models, is increased by unintended accessed points.
- **Detecting threats regarding networks**: arises out of the ubiquitous connectivity issues in different equipment and software. They are likely to spread quickly to the complete network.
- **Advanced metering management techniques**: presently, most of the management schemes are only for facilitating a secure communication within the grid, to address establishment issues of communicating products within SCADA. All in all, only a few studies are carried out on the AMI key management schemes.

11.11.2 COMMUNICATION CHALLENGES

- **Delay of deals and real-time capacity**: Bidirectional information flow among NANs and HANs should meet real-time prerequisites. The information transmissions include numerous kinds of information, which have various degrees of time necessities
- **Self-configuration of AMI**: HANs interface several savvy gadgets to accomplish ideal vitality utilization and to actualize demand responses. Smart meter, smart devices, and EMSs are downloaded at all client premises, all are individual parts of AMI. The AMI will empower these shrewd gadgets to speak with the utility worked control focuses to control their activities at an input time and accordingly actualize demand side management executives for the utility
- **Limited WAN coverage area due to the use of ISM bands**: The communications between NANs and WANs are built up based on cognitive base stations. Hence, there is also the problem of a shortage of licensed bands for opportunistic access.
- WAN connections' feature of scalability using the strange communicating technology is restricted because of quite higher installing and maintaining costs. Therefore, for WAN, wireless communication technologies are accurate.

11.12 SOLUTIONS IN ADVANCED METERING INFRASTRUCTURE

11.12.1 SECURITY SOLUTIONS

- **Quantum distribution key**: The utilization of quantum key distribution (QKD) helps enhance communication securities. It is proposed as a way of improving power grid communication security. It can be actualized over present channels of fiber optic and free space optical communication links, inside generating frameworks and distribution systems. It operates on entangled quantum states physics as a fundamental asset. The old-style cyber security procedures rely upon physical assurance of communication channels, and they require some complex computational methods to encode communicated information and ensure its secrecy. QKD has quickly developed and is currently used in commercial applications by a few organizations around the globe. Analysts are investigating its applications in more challenging and intriguing situations.

- **Attack detecting cross layer designs**: based on CR innovation. Advanced Metering Infra. security incorporates the insurance of both communication systems and power grids so as to guarantee accessibility and survivability.

11.12.2 COMMUNICATION SOLUTIONS

- **Expansion of WAN coverage by hybrid spectrum accessing technique**: since spectrum holes of the authorized bands might not be sufficient to send an enormous measure of information, the HANs and NANs communications may work in license free band incidentally, with decreased rates of communication. In this method, information transmissions among HGWs and NGW are viewed as utilizing Hybrid spectrum accessing. Thus, communication among HANs and NANs might enhance dependability.

11.13 CONCLUSIONS

- This report reviews and presents the complete discussion of smart meters along with their functions and operating requirements. We have also discussed the basic operating principles of the AMI, the smart meters, and also of the MDMS.
- The data structures and data communication's standards, ongoing problems and how AMI is going to remove them are discussed.
- Many several advanced metering's components along with their requirements and function are shown.
- Advanced metering system is the smart grid's most basic information platform and acts as a bridge connecting users to the electric power system. It rapidly measures, along with diagnosing and also adjusting the power qualities, also it monitors and controls the distributed power generation and energy storing devices.
- It also provides intelligent control, is interactive, open and friendly, has high efficiency and is the most economical application of advance metering, that is going to have an advantage of good developments for the construction of smart grids in the coming future.
- The kinds of communication networks that are employed in Advanced Metering are discussed here. Here a multi-communication-based AMI technology has also been proposed and the laboratory set up and field experiments are presented and findings discussed.
- With the implementation of smart meters and advanced metering systems, there exists a significant potential for shifting the electric power supplies.

11.14 FUTURE-DAY SCOPE

- Future smart grids must include smart monitoring systems so as to have a track of the complete electricity flow and also a large quantity of data must be collected out of the smart devices. Therefore, it should be flexible for accommodating new demands in a manner that is economical. In order to achieve this goal, AMI communication based on CR is expected to play a significant role in smart grid infrastructure. Also, smart grids may support real timing traffic delivery, by using AMI.

- In future, AMI architecture not just aims at the seamless integration of several present smart metering products, but also on the software systems undertaken by power utilities, e.g., power outage, distribution, and EMS. Therefore, the aim of the several new alternatives is simply to enable flexible integration of metering devices and to group them as a virtual meter with the help of hierarchically arranged software systems and standardizing communication interface flexibly.

- Future work involves development of fresh algorithms for finding single line to ground fault on Transmission and Distribution systems. Comparative studying of three parameters would be conducted. Those are the distances obtained from real time data, Taurus kit, and from the displayed distance on the smart meters.

REFERENCES

1. Janaka B. Ekanayake, Nick Jenkins, Kithsiri Liyanage, Jianzhong Wu, Akihiko Yokoyama, "Smart Grid: Technology and Application" April 2012 320 Pages.
2. "Smart grid handbook", Indian Smart Grid TechNov, 2017.
3. "Requirement of advanced metering", CEA, Aug. 2016.
4. Gellings, C. The concept of demand-side management for electric utilities. Proc. IEEE 1985, 73,. 1468–1470
5. Clean technica: Dye-sensitized solar cells achieve record efficiency of 15%. Available at: http://cleantechnica.com/2013/07/15/dye-sensitized-solar-cells-achieve-record-efficiency-of-15/ (accessed 19 December 2015).
6. G. Barbose, N. Darghouth, R. Wiser, *Tracking the Sun IV: An Historical Summary of the Installed Cost of Photovoltaics in the United States from 1998–2010*, Berkeley, CA: Ernest Orlando Lawrence Berkeley National Laboratory, 2011. Available at: http://eetd. lbl.gov/ea/emp/reports/lbnl-5047e.pdf (accessed 18 December 2015).
7. C. Becker, T. Sontheimer, S. Steffens, Polycrystalline silicon thin films by high-rate electronbeam evaporation for photovoltaic applications – Influence of substrate texture and temperature, *Energy Procedia* 10: 61–65, 2011.
8. K.W. Boer, Cadmium sulfide enhances solar cell efficiency, *Energy Conversion and Management* 52: 426–430, 2011.
9. M. Boutchich, J. Alvarez, D. Diouf, Amorphous silicon diamond based hetero junctions with high rectification ratio, *Journal of Non-Crystalline Solids* 358: 2110–2113, 2012.
10. J. Britt, C. Ferekides, Thin film CdS/CdTe solar cell with 15.8% efficiency, *Applied Physics Letter* 62: 2851–2852, 1993.

Index